T0192388

Practical Machine Learning with AWS

Process, Build, Deploy, and Productionize Your Models Using AWS

Himanshu Singh

Apress®

Practical Machine Learning with AWS

Himanshu Singh
ALLAHABAD, Uttar Pradesh, India

ISBN-13 (pbk): 978-1-4842-6221-4   ISBN-13 (electronic): 978-1-4842-6222-1
https://doi.org/10.1007/978-1-4842-6222-1

Managing Director, Apress Media LLC: Welmoed Spahr
Acquisitions Editor: Celestin Suresh John
Development Editor: Laura Berendson
Coordinating Editor: Aditee Mirashi

Cover designed by eStudioCalamar

Cover image designed by Freepik (www.freepik.com)

Distributed to the book trade worldwide by Springer Science+Business Media New York, 1 New York Plaza, Suite 4600, New York, NY 10004-1562, USA. Phone 1-800-SPRINGER, fax (201) 348-4505, e-mail orders-ny@ springer-sbm.com, or visit www.springeronline.com. Apress Media, LLC is a California LLC and the sole member (owner) is Springer Science + Business Media Finance Inc (SSBM Finance Inc). SSBM Finance Inc is a **Delaware** corporation.

For information on translations, please e-mail booktranslations@springernature.com; for reprint, paperback, or audio rights, please e-mail bookpermissions@springernature.com.

Apress titles may be purchased in bulk for academic, corporate, or promotional use. eBook versions and licenses are also available for most titles. For more information, reference our Print and eBook Bulk Sales web page at www.apress.com/bulk-sales.

Any source code or other supplementary material referenced by the author in this book is available to readers on GitHub via the book's product page, located at www.apress.com/978-1-4842-6221-4. For more detailed information, please visit www.apress.com/source-code.

Printed on acid-free paper

Table of Contents

About the Author

 **Himanshu Singh** is a technology lead and senior NLP engineer at Legato Healthcare (an Anthem company). He has seven years of experience in the AI industry, primarily in computer vision and natural language processing. He has authored three books on machine learning. He has an MBA from Narsee Monjee Institute of Management Studies, and a postgraduate diploma in applied statistics.

About the Technical Reviewer

 **Anindita Basak** is a cloud architect and DevOps engineer. With more than a decade of experience, she helps enterprises to enable their digital transformation journey empowered with multicloud, DevOps, advanced analytics, and AI. She co-authored the books *Stream Analytics with Microsoft Azure* and *Hands-on Azure Machine Learning* and was a technical reviewer of seven books on Azure along with two video courses on Azure data analytics. She has also worked extensively with AWS Infra, DevOps, and analytics.

Acknowledgments

I'd like to thank my parents and brother for their unbounded support and the Apress-Springer team.

Introduction

This book is structured into three parts. The first part of the book covers the concepts of cloud computing and gives an overview of how AWS works. The second part of the book takes on AWS in detail and covers SageMaker, Step Functions, S3 buckets, ECR, etc. The last part talks about the use cases for AWS services. Different services such as Amazon Comprehend and Extract are discussed here.

Specifically, Part I starts by covering cloud terminologies. It helps you understand the cloud concepts required to use AWS. Then the book discusses the various AWS services that Amazon provides and how they help users in different ways. It discusses the different functionalities of AWS that are categorized under storage-based, compute-based, security-based, etc. By end of the chapters in this part, you will have an overview of how AWS works.

Part II discusses SageMaker in detail. The part starts by running a basic preprocessing script in SageMaker and ends with building a complete end-to-end pipeline of machine learning in it. It covers how SageMaker talks with different services such as ECR, S3, Step Functions, etc., to build the final model.

Part III discusses three use cases of machine learning using some of the other services of AWS. The book discusses how to extract text using Amazon Textract, how to use Amazon Comprehend, and how to make a time-series model using Amazon Forecast.

This book was written to give people who know Python and machine learning some experience with AWS. It teaches you how to use the power of AWS to build your heavy models and how AWS provides you with services to make super models or deploy your custom code with the same AWS support.

PART I

Introduction to Amazon Web Services

CHAPTER 1

Cloud Computing and AWS

This chapter covers the different components of cloud computing and of Amazon Web Services (AWS). After reading the chapter, you'll understand the different important components of AWS, which will make it easier to understand the machine learning components of AWS.

What Is the Cloud?

So, what is the cloud? If you look at memes shared across the internet, you might think the cloud is nothing but someone else's computer that you can use from your own computing device, for your own personal use. Then the question arises, why do we need the other computer when we have our own? It's because our computer may not have things that the other system has. Maybe your budget when buying a system was less than the other person's, and he therefore has more computational power to use. So, instead of buying a new system with more computational power, you can just access the other system for some amount of time and then return to your own system. This is the benefit that the cloud provides. And, by the way, we all know the other system is not just any normal system. Cloud systems are provided by big companies such as Amazon and Google. So even if you are trying to buy a new system with as much computational power as cloud systems, you will not be able to afford it.

Formally speaking, the *cloud* is a particular computing service that is present at a different remote location that we can access using networking or the internet. Cloud services may include storage services, infrastructure services, software services, or any other specific services that you need. Figure 1-1 shows how different devices are connected to cloud systems at a remote location.

© Himanshu Singh 2021
H. Singh, *Practical Machine Learning with AWS*, https://doi.org/10.1007/978-1-4842-6222-1_1

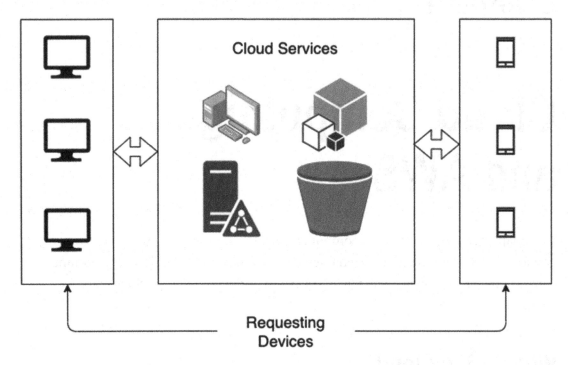

Figure 1-1. *How different devices are connected to cloud systems at a remote location*

If we are able to access any of the services present at the remote location using the internet or networking, then we call this *cloud computing*.

Control of Cloud Systems

Obviously, if someone is allowing access to use their personal system over the internet, then they may want to restrict access in some ways. Or, they may want one group of people to have full access, but another group to have limited access. This is done to avoid security issues and not expose the vulnerabilities present in the system. To solve this problem, cloud computing comes in four types.

- Public cloud
- Private cloud
- Hybrid cloud
- Community cloud

Public Cloud

When the entire cloud infrastructure is open for public consumption, then it is called a *public cloud*. Examples are the email services provided by Google or Yahoo.

Private Cloud

When only a specific group of people can access the services provided by a cloud, then it is called a *private cloud*. An example is when people in an organization can access the resources present in the organization's cloud, but no one from outside the organization can access the same resources.

Community Cloud

When a cloud service is accessible to a group of organizations, then it is called a *community cloud*. For example, different organizations can access the services of AWS or the Google Cloud Platform by registering. So, the same services are available to all the organizations that have paid for it, but not to anyone else.

Hybrid Cloud

When a cloud service provides both options (i.e., services of a public and private cloud), then it is termed a *hybrid cloud*. An example is using two services of AWS. When we train a model using SageMaker training, it is a private task for specific organizations because it contains sensitive data and other things, but when we train a model and then share the endpoint publicly, it is a public cloud because whoever has the link can access that endpoint. (You'll learn more about SageMaker in later chapters.)

Cloud Services

Now that you have learned about the different ways of accessing a cloud, let's dive deeper and look at the services that a cloud platform may provide. We can group these services into four domains.

- Infrastructure as a service (IaaS)
- Platform as a service (PaaS)
- Software as a service (SaaS)
- Anything as a service (XaaS)

Infrastructure as a Service

As the name suggests, when a cloud service provider gives access to users to the infrastructure that it has built, it is considered an IaaS. For example, a cloud provider may give access to virtual machines, physical machines, storage devices, etc. For example, we can use Google Drive to store information on the cloud, since Google is providing its hard drives as a service. AWS also provides machines called EC2 instances that individuals can use to do operations that require higher computational power.

Platform as a Service

Sometimes, instead of requiring an entire infrastructure, we want only a specific development runtime where we can write our code or make games or websites. This way the cost of building an entire infrastructure can be reduced. This type of service is a PaaS. For example, we can use Google Colaboratory for writing Python or R code. In addition, we can use AWS SageMaker to train and put a machine learning model into production. There are other service providers as well such as Microsoft Azure, Google Cloud Platform, IBM Cloud, etc.

Software as a Service

When we don't want the runtime, but we want to use a specific software application with its built-in runtime, we don't need a PaaS, which would give us the runtime as well as dependencies and software we'd need to install. Hence, there are cloud services that provide specific software for specific uses, called SaaS. Examples of SaaS are Amazon Ground Truth, which is used for data management, and Office 365 by Microsoft.

Anything as a Service

The first three types of services have been on the market for quite some time, but now, because of the advancement in technology, cloud service providers are providing almost anything as a cloud service. For example, we can now draw sketches of web pages and give them to Azure, which converts them into HTML pages. In addition, you can play online songs by just talking to Alexa, which is connected to AWS. All this comes under the umbrella of XaaS.

Let's now dive deeper into a specific cloud service provider, called Amazon Web Services (AWS).

Introduction to Amazon Web Services

AWS provides global cloud computing services across many countries and is currently responsible for handling the infrastructure of many companies, including small and large enterprises. According to the AWS documentation, currently AWS caters to hundreds of thousands of businesses in 190 countries.

AWS provides more than 150 services that can be used on demand and can be paid for based on the time used. Currently, AWS has data servers in a lot of regions, and you can choose to use only one region's server that is closest to your users. The following is the list of data servers across the globe:

- **North America**

    - Ohio (US East)

    - Oregon (US West)

    - Northern California (US West)

    - Northern Virginia (US West)

    - Gov Cloud (US East and US West)

    - Canada (Central)

- **South America**

    - Sao Paulo

- **Europe/Middle East/Africa**

    - London (Europe)

    - Stockholm (Europe)

    - Frankfurt (Europe)

    - Paris (Europe)

    - Bahrain (Middle East)

    - Ireland (Europe)

- **Asia Pacific**

  - Singapore (Asia Pacific)

  - Beijing (Mainland China)

  - Sydney (Asia Pacific)

  - Tokyo (Asia Pacific)

  - Seoul (Asia Pacific)

  - Ningxia (Mainland China)

  - Osaka (Asia Pacific)

  - Mumbai (Asia Pacific)

  - Hong Kong (Asia Pacific)

As I mentioned, AWS has more than 150 services. The question is, how do you access them? Is there a single centralized place from where they can be accessed? Well, yes! This place is called the AWS Management Console. Let's look at some of the features of this console and how it is really helpful to users.

AWS Management Console

With the AWS Management Console (AMC), not only can you access the services, but it provides some other cool features as well. Some of them are as follows:

- Once you have created an account on AWS and logged in to AMS, then your session remains active only for 12 hours. After that, you need to log in again. Obviously, this time limit is customizable. This feature is provided for security reasons.

- Not only can you access AMS from the Web, but you can use the mobile app as well. The AMS app is present both on IOS and on Android devices.

- AMS provides access to different learning resources, articles, documentation, videos, etc., which help us in understanding the different services of AWS.

- You can even customize and personalize AMS based on your usage and needs.

After logging in to AWS, you will see the following features:

- Search button to find specific services

- Recently visited services by a user

- List of all the services

- Links to automated workflows

- Link to learning resources

Figures 1-2, 1-3, 1-4, and 1-5 show the different screens of AMS.

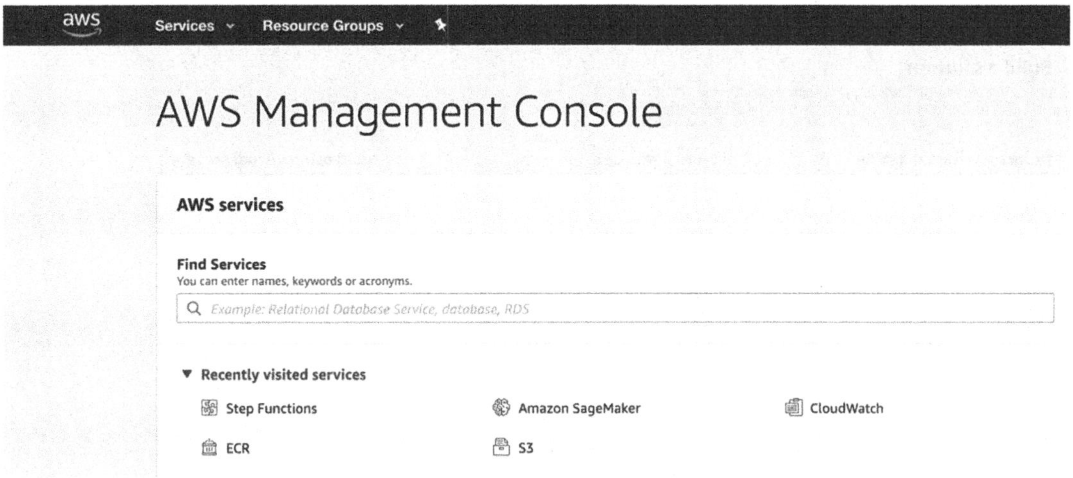

Figure 1-2. *Find Services feature and recently visited services on AMS*

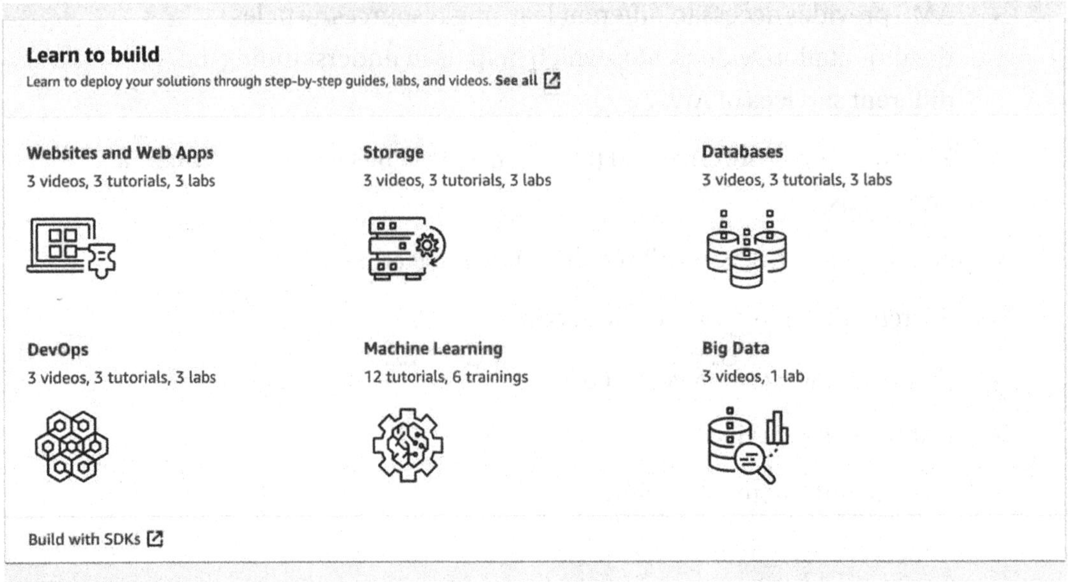

Learn to build

Learn to deploy your solutions through step-by-step guides, labs, and videos. **See all** ☑

Websites and Web Apps

3 videos, 3 tutorials, 3 labs

Storage

3 videos, 3 tutorials, 3 labs

Databases

3 videos, 3 tutorials, 3 labs

DevOps

3 videos, 3 tutorials, 3 labs

Machine Learning

12 tutorials, 6 trainings

Big Data

3 videos, 1 lab

Build with SDKs ☑

Figure 1-3. *Learning resources on AMS*

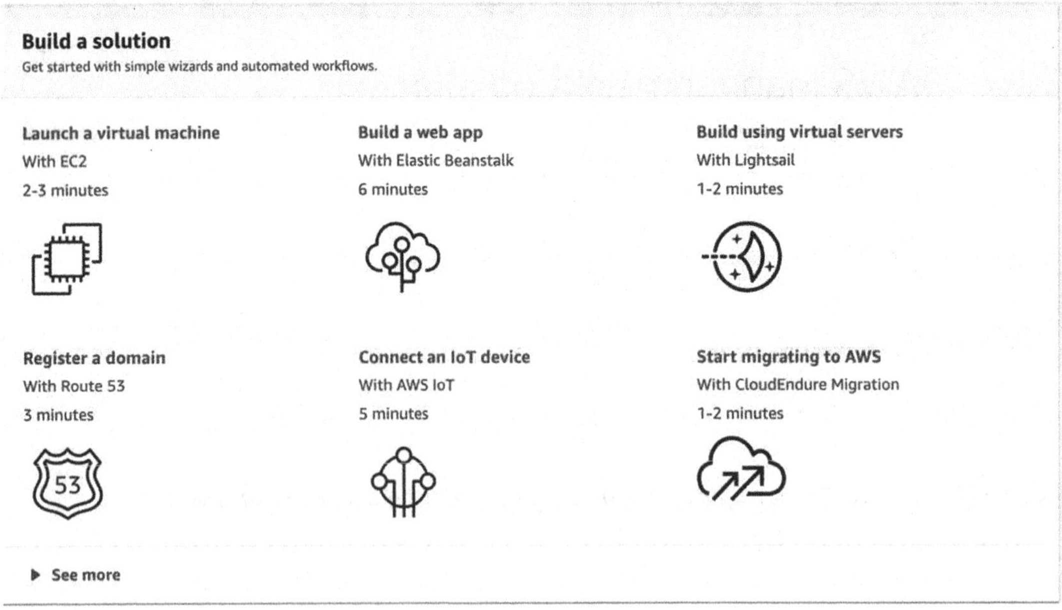

Build a solution

Get started with simple wizards and automated workflows.

Launch a virtual machine

With EC2

2-3 minutes

Build a web app

With Elastic Beanstalk

6 minutes

Build using virtual servers

With Lightsail

1-2 minutes

Register a domain

With Route 53

3 minutes

Connect an IoT device

With AWS IoT

5 minutes

Start migrating to AWS

With CloudEndure Migration

1-2 minutes

▶ See more

Figure 1-4. *Automation on AMS*

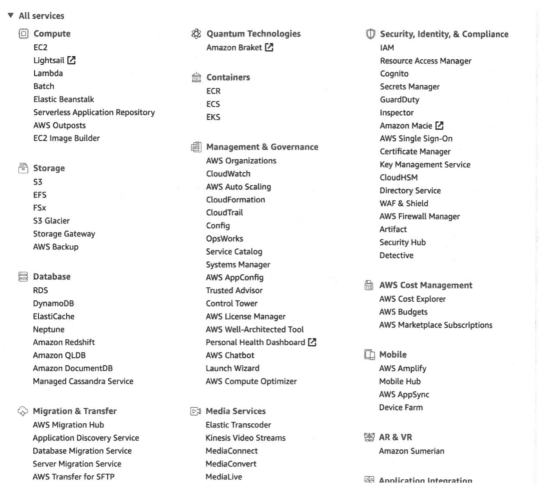

Figure 1-5. *List of all services on AMS*

We will look at how to log in to AWS and visit AMS in detail in the next section about machine learning. Now, let's move to the next feature of AWS called the AWS Command-Line Interface (AWS CLI).

AWS Command-Line Interface

If AMC gives you a visual interface to access the AWS services, the CLI gives you some advanced power to access the same services through the console. It is used by advanced developers who have spent some time with AWS. You just need to download a single tool, and then you can use it to control different services, write scripts, and have control over the automation of services.

AWS provides a lot of resources such as reference documents, GitHub repositories, forums, etc., for understanding AWS CLI. Though one can use AWS CLI from the default console such as the command prompt of Windows or terminals of Linux and Mac, there is a dedicated AWS shell that provides some advanced functionalities. Some of them are as follows:

- Autocompletion support

- Inline documentation of commands

- OS shell commands, which can also be executed from the same shell

We will be using the CLI a lot when we will cover machine learning in detail. Therefore, we will look at its practical aspects directly in that section.

Because AWS provides so many services, covering all of them is not possible in one book. Moreover, this book is about machine learning, so it doesn't make any sense to cover every service here. But, we will discuss three services that I think are really important and commonly used. The following are the services that we are going to discuss here:

- AWS Storage Services

- AWS Compute Services

- AWS Networking and Content Delivery Services

Let's start the discussion with the first one, Storage Services.

AWS Storage Services

When we work on a cloud platform and use its services, obviously we'll have a lot of data depending on the requirements. For example, if we are building a website, then we will have images, videos, and lots of other things to store. If we have a machine learning model, then we will have terabytes of data to handle. This data can be both structured and unstructured. Similarly, for business purposes, we can have multiple Excel sheets or presentations. All these data types must be stored somewhere in the cloud, and the cloud platform should provide this facility.

AWS provides a lot of options for data storage, and we'll discuss three of them in this section.

- Amazon S3

- Amazon Elastic File System (EFS)

- AWS Storage Gateway

Amazon S3

One of the most used services of AWS is Amazon Simple Storage System (S3). It provides you with an interface where you can store your data in a similar way to how you store it in your local file system. You can create folders and multiple subdirectories to organize your data. The following are some of the basic features Amazon S3 provides:

- It provides scalability, which is currently leading in the industry.

- It provides real-time data availability.

- It provides security and optimized performance.

- It has a durability of 99.9999999... percent (11 nines).

S3 is really simple to use. First let's understand some of the naming conventions used by Amazon S3.

Buckets

A *bucket* is just like a folder in your local file system. It is a container used for storing your files.

Objects

The files that you store in S3 are termed *objects*. All the objects are stored inside the buckets.

Keys

Every object that you store will be given a unique identifier called a *key*. Also, not only objects but buckets are provided with unique keys.

Does S3 only provide simple storage, as its name suggests? I will say yes and no. Yes, because its main use is storage only, and it is really simple. No, because it has lots of other features revolving around the storage feature that make it a go-to service for every customer. Let's see what those features are that make S3 so powerful.

- Based on how frequently data is being used, S3 provides different types of storage classes.

    - **S3 STANDARD**: Data that needs to be frequently accessed

    - **S3 STANDARD_IA**: Data that needs to be less frequently accessed

    - **S3 GLACIER**: Data that we want to archive

- Storage without security is nothing. AWS provides access control to the buckets that you have created. You can accomplish this using *policies*. The following are the three levels of control based on policies that we can apply:

    - Who can access which bucket?

    - From which network can the buckets be accessed?

    - At what time should the buckets be accessed?

- You can also create versions of your objects. For example, if the same Excel sheet is updated five times, then five versions of it can be created.

In this entire book, Amazon S3 is the service that we will be using continuously with machine learning services. We will discuss the services in detail in the next section.

Amazon Elastic File System

Amazon Elastic File System (EFS) is an elastic network file system that most of the AWS cloud services are compatible with. It is called *elastic* because it is scalable as well as shrinkable. If you upload a smaller amount of data, then it shrinks its size to accommodate that data. But if you upload a larger amount of data, then it can scale up its size. Scaling up can be in the petabytes as well. EFS works with the latest version of NFS, which is NFSv4.1. Hence, it is compatible with almost everything that you want to develop.

Tip   Using Network File System (NFS), you can store, edit, delete, and perform other operations similar to how you perform them in your local system. It is a kind of distributed file system that uses *network-attached storage* (NAS). The current version of NFS provides advanced features such as strong authentication, file caching, and support for Windows File System. NFS can be accessed now on global WANs.

Just like S3, EFS provides two kinds of file storage.

- Standard Access
- Infrequent Access

When we want to access data frequently, we use Standard Access, while infrequently used data can be stored in Infrequent Access EFS. Also, just like S3, you can authenticate and authorize data in EFS and encrypt it further. Finally, you can add policies, just like S3, for maintaining access control.

AWS Storage Gateway

AWS Storage Gateway is a hybrid infrastructure provided by AWS. If you want to use your on-premise infrastructure for all your storage needs but still you want some functionality by which you can use the cloud storage services of AWS, then Storage Gateway is the best solution.

Storage Gateway provides three kinds of solutions.

- File Gateway
- Volume Gateway
- Tape Gateway

File Gateway

Using this service, all the files are stored in S3. It gives you a virtual application with which you can manage all your files in S3. Retrieving/storing files is done using protocols such as Network File System or Server Message Block. The virtual software that we are talking about is nothing but a virtual machine with which you manage your files. This can be with VMware ESXi or Microsoft Hyper-V.

Volume Gateway

Instead of files, you can directly store volumes in the cloud that you can later mount as Internet Small Computer System Interface (iSCSI). Again, the software that is deployed on-premise is a virtual machine. The following kinds of volumes are supported:

- Cached volumes

- Stored volumes

Having cached volumes means storing the data entirely in S3, and then the frequently used data is cached in the local system. Figure 1-6 shows the cached volume gateway architecture provided by AWS.

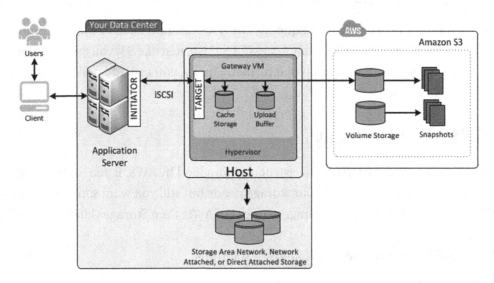

Figure 1-6. *Cached volumes, Storage Gateway architecture*

Figure 1-6 is divided into three parts. The left part shows the actual users using the local architecture. The middle component is the local infrastructure of an organization. The right component has an S3 connection for the data backup.

When you store your entire data locally and then back up the snapshot versions of this data on the cloud, then it is the stored volume support of Volume Gateway. We can use this in the case of disaster recovery. For example, if you lose your local data, you can download the latest snapshot from the cloud. Again, we use S3 as the storage service here. Figure 1-7 shows the architecture of a storage volume.

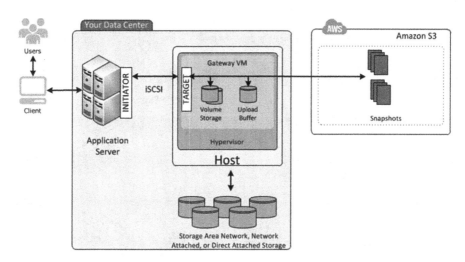

Figure 1-7. *Stored volumes, Storage Gateway architecture*

The architecture in Figure 1-7 is almost the same as Figure 1-6, but instead of storing the entire data, we are storing only snapshots of the locally saved data.

Tape Gateway

This is used for archiving data. For this we can use Amazon S3 Glacier or Deep Archive as the storage service. This can also be deployed locally using virtual machines. The architecture given in Figure 1-8 shows how tape gateway works.

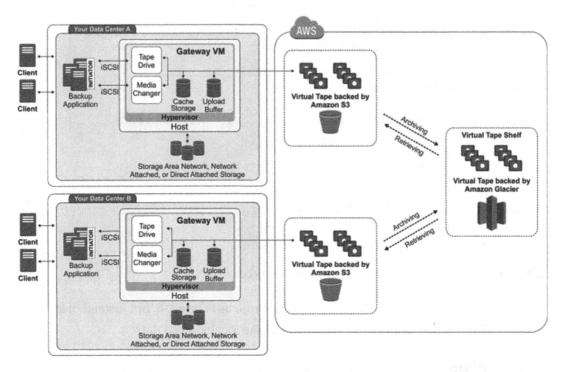

Figure 1-8. *Tape gateway, Storage Gateway architecture*

You can see in the architecture that different infrastructures are storing the data in their respective S3 buckets. Later, the data of all the S3 buckets is combined and then stored in the Amazon S3 Glacier. Virtual tapes are nothing but a means of storing data. Just like how physical tapes were empty and then filled at the time when they were used, similarly virtual tapes can also be blank and can be filled with data as per your needs.

Now that we have seen how Storage Services of AWS operates and looked at different services, let's move on and explore the compute services.

AWS Compute Services

In the previous section, you learned how to store data using AWS services. In this section, we will look at some compute services provided by AWS. When we run an application, play games, or develop something, we require the computational power of the system. We measure this in terms of the RAM, processor, graphics card, etc. Sometimes we may have a big system requirement but not be able to get it due to lack of money or lack of resources. That is the reason why Amazon provides different kinds of services for all computational requirements. In this section, we will discuss the following AWS compute services:

- Amazon EC2

- Amazon ECR

Amazon EC2

Elastic Component Cloud (EC2) reduces the burden for a user to invest in hardware requirements. Whatever the requirement, EC2 allows you to create that many virtual servers where immediately the work can be processed. In addition, instead of having a static infrastructure, it is dynamic. This means that if a sudden surge in computation requirements occurs, EC2 automatically scales itself up, without disrupting the ongoing processes.

The following are some of the features that EC2 gives its users:

- It provides an environment where we can do our computation-heavy work. This is done in a virtual computing environment with what AWS calls *instances*.

- Whatever your software or hardware needs, you can configured them using the service Amazon Machine Images (AMIs).

- It provides security to all the instances you spin up. It also provides security groups that help the user configure firewalls, ports, IP ranges, etc.

In addition, there are a lot of other services that help the users in their day-to-day coding and development life.

The instance types that EC2 provides can be grouped into the following categories:

- General purpose

- Compute optimized

- Memory optimized

- Storage optimized

- Accelerated computing

General-Purpose Types

These types provide a balance of computational power, storage, and networking. They are further divided into the following groups:

- A1
- T3
- T2
- M6g
- M5
- M5a
- M5n
- M4

Compute Optimized

When the work requires high computational usage and requires heavy processors, you can use compute-optimized instances. They can be used for media tasks, scientific tasks, fast web servers, or even game servers. They can be grouped into the following types:

- C5
- C5n
- C4

You can find more information about these groups at `https://aws.amazon.com/ec2/instance-types/`.

Memory Optimized

If the work requires working with large datasets, you need memory-optimized instances. They are grouped into the following types:

- R
- R5a
- R5n

- R4

- X1e

- X1

- High Memory

- z1d

Accelerated Computing

For all machine learning and deep learning applications, these kinds of instances are preferred. These systems are really fast, and their precision is also very high. They can be grouped into the following types:

- P3

- P2

- Inf1

- G4

- G3

- F1

Storage Optimized

This type is best if you want to work on huge datasets and want less latency with faster read and write operations. They can be grouped into the following types:

- I3

- I3en

- D2

- H1

You can find detailed information about every instance types at https://aws. amazon.com/ec2/instance-types/.

Other Services

AWS also provides other services with most of the previous instance types. This is done to give you financial benefits based on your usage. These services include the following:

- On-demand instances

- Reserved instances

- Spot instances

If you think your work is not going to take a lot of time, like for days or months, then you can go for *on-demand instances*. These instances are charged on a per-hour basis. Whatever compute or memory and storage size you want, you can use that, but the payment will be based on the hours used.

Reserved instances are for longer-term usage, and hence the discount that a person receives is up to 75 percent of the on-demand instances. This also provides options to change types.

There are times when a lot of EC2 instances get unused. *Spot instances* let you take advantage of those instances, and you can get up to a 90 percent discount as compared to on-demand instances when using spot instances.

Amazon Elastic Container Registry

In recent times, Docker has taken the industry by storm. It allows companies to separate their infrastructure from their coding. In simple words, Docker provides you with a platform where you can develop and ship code efficiently without worrying about the underlying architecture. Amazon ECR is a repository for all the Docker images that you want to run in AWS.

Once you spin up an EC2 instance, you have an option to run any Docker image on that instance. AWS has its own prebuilt Docker images that you can import, or you can have your own custom-made Docker images uploaded on ECR and then imported inside the EC2 instance.

Amazon ECR has the following components:

- Registry

- Authentication token

- Repository

- Policy

- Image

The registry is like a normal register, where you want to make an entry to every image that you upload to ECR. To make sure that only the right person is able to access the ECR for uploading images, authentication tokens are used. The repository is the place that actually stores your Docker images. Images are your actual Docker files. These are the files that you have created containing all your dependencies. You use the image to store them in ECR and import them in EC2.

You will understand the operations of ECR in detail when we cover SageMaker in later chapters.

AWS Networking and Content Delivery Services

In this section, we will discuss three important services of AWS in this domain.

- Amazon VPC

- Amazon API Gateway

- Amazon CloudFront

Amazon VPC

Amazon Virtual Private Cloud (VPC) is a virtual network that we create so that we can segregate certain things from the entire user domain. It acts just like a normal cloud, but instead of having separate infrastructures, there is only one cloud infrastructure but multiple virtual clouds made over it. For example, we can make a separate virtual cloud for the marketing, finance, and operations departments. Only its own set of users know what is happening in each cloud, but actually all the files that are being stored are on the same storage, which is being shared by all the departments. Therefore, each cloud can have its own security policies, access levels, etc.

Amazon VPC is the same, except it also provides a scalable AWS architecture. Amazon VPC has the following components:

- *Subnet*: Each virtual private cloud can be accessed only by a set of IP addresses. Any request coming from an address outside the list is not given access. This list is called a *subnet*.

- *Route table*: As I said, a virtual private cloud has the same underlying infrastructure as local cloud. But AWS provides its own features as well. One of the features is load balancing, meaning if the load of the server becomes very high, then we can divert the traffic to an alternate server. Once we are working in VPC, we must know the routes to where the traffic needs to be directed. These routes are stored in route tables.

- *Internet gateway*: Through the Internet gateway, all of your virtual private clouds are able to contact the underlying EC2 instances.

- *Endpoint*: If we want to connect our virtual private cloud to any of the services provided by AWS, we can just use the VPC endpoint service provided by Amazon VPC.

Figure 1-9 shows an architecture where the Internet is used for VPC usage.

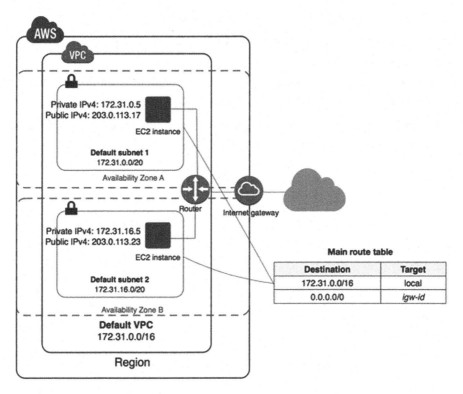

Figure 1-9. *Amazon VPC architecture*

If you look at the architecture in Figure 1-9, you can see that inside VPC we have two EC2 instances, each with a public IP address and a private IP address. The public IP address is used if you want to access the services from outside the VPC, while the private one is used to access the same services from the inside of VPC. You can see that there is a router that uses route tables to know different routes and then uses an Internet gateway for the access. You also have an IP address and a port number for the VPC (172.31.0.0 is the IP address and 16 is the port number in Figure 1-9).

Amazon API Gateway

Before understanding API Gateway, I will give a brief introduction about application programming interfaces (APIs). An API is a service that we use to make two or more applications talk with each other. For example, when we use Facebook to upload an image or a video, we are using the upload API of Facebook, and whenever we are liking, commenting, or sharing, we are using another API of Facebook. Therefore, in the current development scenario, each small service is developed and then converted into an

API (the REST API is one of the types of API that is most used) and then can be used using the networking protocols. Amazon API Gateway provides you with the services to efficiently manage these APIs.

Amazon API Gateway can be used to create, publish, maintain, monitor, and secure different kinds of APIs such as REST, HTTP, or the WebSocket API. These APIs can be made not only to have their own applications, but they can also access and use different AWS services.

Figure 1-10 shows the architecture of API Gateway.

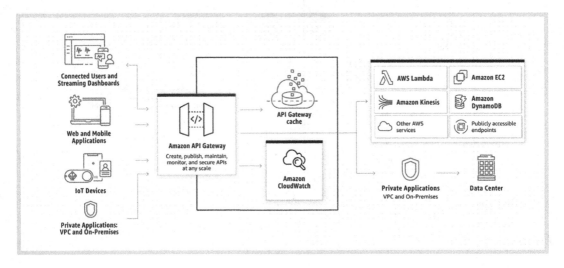

Figure 1-10. *Architecture of API Gateway*

In Figure 1-10, Amazon CloudWatch is used for monitoring the logs of any services provided by AWS and API Gateway in the current state. API Gateway Cache provides you with an option to reduce the latency by storing the most used components. All the APIs can be connected to different services of AWS as well like AWS Lambda, EC2, Kinesis (Used for Live Data Analytics), DynamoDB (Cloud Database), etc. It can be used by any third-party applications as well.

Amazon CloudFront

There are a lot of services present on the Web that provide us with a lot of options to build websites and serve them to the end users. AWS can be counted as one of them, but it provides features that the others don't provide. AWS gives the option of less latency. To understand latency in a little detail, imagine you are opening a website whose servers are in India, but you are in North Korea (assuming that the Internet is provided to its citizens). Now, since the servers are pretty far away, it will take a lot of time to use any service provided by the website. But if, instead, the website was deployed in AWS, then instead of using the servers present in India, a person would be able to use the servers present in South Korea. Because the server is closer to the country, the pages will open faster.

How does AWS do it? The answer is its service called CloudFront. How does CloudFront do that? It does that by using the edge locations provided by AWS. When you make a website, it has both static and dynamic content such as HTML pages, CSS and JavaScript, etc. So, if a request to use the website comes from the location where your server is not present, CloudFront distributes all the contents of the website to the Edge Location nearest to the place from where the request was made. Therefore, now opening the website becomes faster as the content is being delivered from a nearest server. To know more about edge locations, please visit this link: `https://aws.amazon.com/cloudfront/features/`.

In Figure 1-11, you can see how CloudFront is used for this distribution. Let's see what each step represents in the image.

1. All the files that your website is going to use are stored in an AWS service like S3, or it can be your own HTTP server. (Steps 1 and 2.)

2. Now you initiate the CloudFront distribution by providing the link to your S3 or HTTP server. (Step 3.)

3. A domain name, specific to CloudFront, is provided to your distribution. It can be changed as well. (Step 4.)

4. CloudFront sends the configuration of the distribution it has just created to all the edge locations present across the world. Here the cache of your files is created. (Step 5.)

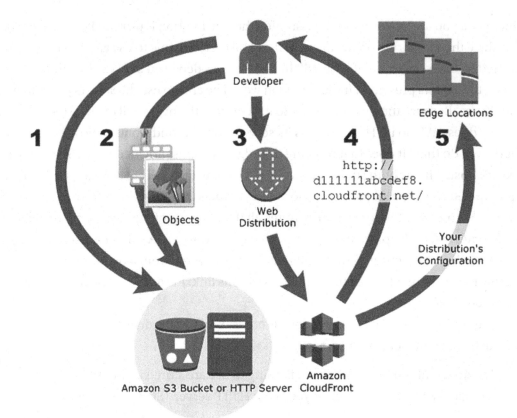

Figure 1-11. *Amazon CloudFront service*

Conclusion

This finishes the basic introduction about cloud computing and the Amazon version of the cloud: AWS. In this chapter, you saw different concepts related to cloud computing and learned about different components of AWS. We have not seen all the services of AWS, as there are more than 150 of them. They all cannot be covered in a single book, but we have covered all the important ones as far as this book is concerned.

In the next chapter, we will be looking at the security aspects of AWS and different types of services provided to make our applications secure.

CHAPTER 2

AWS Pricing and Cost Management

In the previous chapter, we looked at some concepts of cloud computing and explored a few of the important services provided by AWS. In this chapter, we will look at how AWS charges us for the services it provides and how we can get the best out of AWS with the least amount of burden on our pocketbooks as possible.

Understanding the Pricing of AWS

As per the documentation, AWS tries to make our lives simpler by charging us the same way as we get charged for using electricity or water. This means we pay only for whatever we are going to use. When we have an infrastructure present on our premises, we have to pay for everything, even if we are not using it. In AWS, it is not the same. This is called *pay-as-you-go*.

In Figures 2-1 and 2-2, AWS tries to show the difference in pricing between two architectures. When we look at Figure 2-1, we see that even when we are not using the infrastructure, the part denoted by the red shaded region, we are still paying for it. Hence, in the end, you see that every month the charges keep increasing, and by the end of the year you end up paying a lot.

© Himanshu Singh 2021
H. Singh, *Practical Machine Learning with AWS*, https://doi.org/10.1007/978-1-4842-6222-1_2

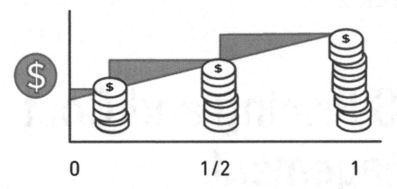

Figure 2-1. *Pricing in on-premise infrastructure*

Instead, Figure 2-2 shows that one pays more if it uses more, while in the remaining days it is much less. Hence, the cost savings in the case of AWS is much more. Another way that AWS helps us to save money is by using the reserve functionality provided.

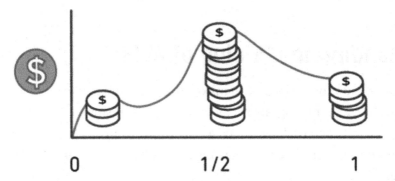

Figure 2-2. *Pricing in AWS infrastructure*

By *reserved instances*, AWS means that you pay for a service of AWS up front, for a longer time of usage. This means the resources and infrastructure required for that service at a later time are reserved, and hence no one else can use those reserved services and infrastructure. As you are asking to use an entire service for a longer amount of time, this means AWS will keep on earning from those services. The infrastructure will never be idle in terms of money. Therefore, to make things easier for the companies opting for reserved instances, AWS charges them less. It is sometimes less than 70 percent of the on-demand instances that we talked about in the previous chapter.

Now, before going into the details of AWS pricing, let's first look at the AWS Free Tier, including what services we can use for free and for how much time.

AWS Free Tier

AWS offers free services in three areas.

- Services that are always free
- Services that are free for around 12 months
- Services that are available on a trial basis

The following are some of the services that are always free:

- DynamoDB is free for 25 GB storage.
- AWS Lambda is always free for 1 million requests per month, and up to 3.2 million of compute time.
- 1 million objects can be stored in Amazon Glue for free.

The following are some of the services that are free for about 12 months:

- 5 GB of Amazon S3
- 750 hours of EC2 instances per month, supporting t2.micro and t3.micro
- 50 K units of text in Amazon Comprehend

The following are the services that are available on a trial basis:

- 250 hours/month of Amazon SageMaker using t2.medium for two months
- 50 hours/month of Amazon SageMaker using m4.xlarge for two months
- 125 hours/month of hosting in Amazon SageMaker for two months

Factors Affecting Pricing in AWS

AWS calculates pricing based upon these three elements:

- Computation
- Storage
- Data transfer

AWS doesn't charge you for the incoming transmission of data or data transmission between two or more services of AWS only. But, whatever data goes outside the AWS network (an outbound data transmission), it does charge you for that. It depends upon the amount of data transferred. The more data, the lower the price per gigabyte.

As mentioned, for computations the charge is calculated on an hourly basis, and for storage it is calculated in gigabytes. Now that we know which factors affect AWS pricing, we must know how we can optimize the cost in AWS.

AWS Cost Optimization

We can create a cost-effective AWS architecture by following the four simple steps given here:

- Right-sizing your services

- Using reserved instances

- Using a spot market

- Using the Cost Explorer

Right-Sizing

The term *right-sizing* means that you only use a service that you need. This means you only use the compute power that you require or only use the storage that you need. You must not over-provision it and neither should you compromise on the capacity. AWS provides you with services that help you in right-sizing through autoscaling, which includes up scaling, down scaling, scaling out, and scaling in based on the usage.

To right-size, the organization should perfectly know the needs and usage pattern required and then take advantage of AWS's elasticity to right-size. Through this, the organization can save up to 70 percent of the total cost. Also, one must remember that it is not a one-time affair. This means the companies have to periodically plan their needs and pattern and make changes accordingly. Therefore, one can say that right-sizing is an ongoing process.

How do you decide when it is the perfect time to perform right sizing again? For this, AWS has given us some tools that can be used for making these sorts of decisions.

- *Amazon CloudWatch*: You can know about the amount of CPU utilized, the throughput of the network, the disk I/O operations, etc., and then use this information to decide whether revised right-sizing is required.

- *AWS Cost Optimization*: This provides you with the recommendations about the right-sizing based on the current utilization. We will discuss this more in the upcoming section.

- *AWS Cost Explorer*: You can use the Cost Explorer to understand what are the prime drivers of the cost incurred to the company. Based on that, the right decisions can be made.

- *AWS Trusted Advisor*: This helps you know more about idle or underutilized resources.

The following are some of the factors that can be taken into consideration while deciding about the optimal right-sizing. These factors can be grouped into the following categories:

- Right-sizing based upon performance data

- Right-sizing based upon usage needs

- Right-sizing by stopping instances

- Right-sizing based upon selection of right instance family

- Right-sizing based upon selection of database instances

When it comes to the analysis of performance data, one should identify the instances that are idle or are underutilized. For this, one can look at the CPU and memory usage of the instances. This can be done using CloudWatch or any other tools mentioned earlier. Amazon recommends that those instances with a maximum CPU or memory usage of less than 40 percent for a four-week period can be right-sized.

When it comes to analyzing the usage needs, one must take into consideration the following:

- If the load remains constant for a longer period of time, we can opt for reserved instances, instead of on-demand or spot instances. We will talk about reserved instances in detail in the next section.

- If the load is not constant, but we can say that in approximately equal intervals or during specific scenarios the load increases or decreases, then we can use the autoscaling features of AWS EC2 instances.

- All loads that are considered to be flexible, which means they are used only when required and then can be turned off, can use on-demand or spot instances.

When we run our normal systems such as laptops or desktops, we turn them off when not in use. The same thing can be done with AWS. All those instances that have been idle for more than two weeks can be stopped or terminated. Once an instance is stopped, the company stops paying for the EC2 instance. But remember, for every EC2 instance there is an EBS volume associated with it. The volume remains alive, and you still keep paying for it. If you want to terminate the instance, the attached EBS volume gets deleted as well; hence, there is no cost for EBS. But, if you want to rerun that instance, then some re-provisioning should be done to get back the EBS volume. One of the best ways can be to store the snapshot of EBS volumes so that during termination or even during deletion, the EBS volumes can be stored in the form of snapshots as a backup.

What Is an EBS Volume?

Before understanding Amazon Elastic Block Storage (EBS), we must know what a block device is. A *block device*, in simple terms, is a device that is used to store your information. Therefore, all the disk drives like HDD or SSD are the block devices. They are platform independent, which means you can give them any operating system, and a block device will work. Inside a block device all your files are stored and which can be accessed using any instance and operating system.

Now, when we talk about EBS, it allows this block storage in Amazon EC2 instances. It provides you with the feature to attach multiple EBS volumes in one single EC2 instance or one EBS volume to multiple EC2 instances.

We can decide to select the perfect instance for our workloads, and we can also change it based on two options: changing the instance in the same family and changing the instance to a different family. We can change the instance to another instance in the same family based on the following metrics:

- Count of vCPUs

- Looking at the memory

- Looking at network throughput

- Looking at the attached storage

But, when we change the instance to another instance in a different family, we consider the following metrics:

- Selecting the right virtualization type

- Selecting whether you need VPC support

- Selecting the right platform

- Selecting whether to upgrade hardware requirements or not

For the first three points, we must be sure that the configuration before the upgrade is the same as after the upgrade. For the last point, it must require efforts to move the entire architecture to an upgraded EC2 instance. We talked about different families of EC2 instances in the previous chapter.

What Are Virtualization Types?

Linux machines in AWS support two types of virtualization: paravirtual (PV) and hardware virtual machine (HVM). The difference between these two instances is in how the operating system boots and how it takes advantage of additional configurations like CPUs, storage, etc.

Right-Sizing Database Instances

Last but not the least, we can right-size the database instances. For this, the following metrics must be kept in mind:

- Either you scale up or you scale down your database instances, or the storage size remains the same.

- We can separately change the storage size of the database instance.

- AWS takes licenses seriously. Therefore, whatever database a person is using, the right licenses must be there.

- You can select the right-sizing time of your database instances. It can be done immediately or at a specific window like during the maintenance time.

This concludes our discussion about right-sizing in AWS. Next, we must learn more about reserved instances.

Using Reserved Instances

Reserved instances are all about commitment. While you are purchasing the reserve compatible services of AWS, at that time only must you decide whether you want to go for a long-term commitment or your requirements will change periodically. If it is long-term, then you can opt for reserved services, and this can lead to a reduction in hourly cost of that service. The following are the services provided by AWS that offer a reserve facility:

- Amazon EC2

- Amazon RDS

- Amazon ElastiCache

- Amazon RedShift

- Amazon DynamoDB

One must be absolutely sure to use reserved instances, as the lock-in period is either one year or three years. That means you must pay for either the entire one year or three years beforehand. After the payment is done, no matter how much you use the instance, you will not get any money back, like on-demand instances. Therefore, once you pay for the lock-in period, you can get a discount up to 75% of that of on-demand instances, but then you will not get dynamic pricing for the usage patterns.

Reserved instances come up with the following payment options:

- No up-front payment

- Partial up-front payment

- All up-front payment

As the names suggest, no up-front payment means you don't pay any amount at the start. Partial up-front payment means you may pay some amount at the start, while all up-front payment means that you pay the complete amount at the start. For the no up-front scenario, the customers are charged at the discounted hourly rate. But, since AWS is not sure whether customers are going to pay, it requires a contractual agreement with them, and it also looks at the past relationship of those customers with AWS. For partial up-front payment, the same thing is followed. The remaining amount is charged at a discounted hourly rate.

Reserved instances come under two offerings.

- Standard reserved instances

- Convertible reserved instances

With standard reserved instances, the instances can be increased in size or availability zones can be modified, or they can also be sold in the reserved instances marketplace. Convertible reserved instances, on the other hand, can be exchanged with other convertible instances that may have new attributes such as instance family, instance type, platform, etc. They cannot be sold in the marketplace.

There is also a limit on the number of reserved instances that can be purchased in a month. For a particular region, 20 regional instances and 20 zonal instances can be purchased in a month. A regional instance is an instance that is available for a complete region, while a zonal instance is an instance available to a specific availability zone.

Using Spot Instances

As already discussed in the previous chapter, spot instances are those instances that are running idle and currently AWS is not generating any money from it. These instances can be taken by companies and can achieve discounts up to 90 percent of the on-demand instances. Remember, as these are the instances that are currently idle, that doesn't mean they will always be idle. This means that whenever the use of them goes up again, EC2 gives you a two-minute notice and then interrupts the session of your spot instance. Obviously, every great thing has disadvantages. Another way by which your spot instance can be terminated is when the cost incurred by you increases the threshold defined.

We must know about the scenarios in which one must use the spot instances. Amazon recommends using the spot instances for the following scenarios:

- Making fault-tolerant and flexible applications

  - Web servers

  - API back ends

  - CI/CD pipeline (DevOps)

  - Hadoop data processing

- Image rendering

- Stateless web services

- Big data analytics

- Parallel computations

Once you have an on-demand instance, you can ask for spot instances to handle some extra functionalities that your application has. For this you can request spot instances by launching a wizard in EC2. You just need to tell the number of instances, the type of instances, the availability zone, and the maximum price that you're willing to pay for the same.

Using the Cost Explorer

Once you have decided on the type of service to use, whether reserved, on-demand, or spot, you can analyze your usage and costs incurred using a tool provided by AWS called the Cost Explorer. There are three types of reports that the Cost Explorer provides to help you: reporting on the usage and cost incurred in the past 12 months, forecasting how much you are going to spend in the next three months based on your past usage, and getting recommendations as well on the type of instances to use.

If you open the Cost Explorer dashboard, it will come with a default view, as shown in Figure 2-3.

Figure 2-3. *AWS Cost Explorer dashboard*

Looking at the dashboard, we can say that a Cost Explorer consists of the following components:

- Cost Explorer Costs

- Cost Explorer Trends

- Daily Unblended Costs

- Monthly Unblended Costs

- Net Unblended Costs

- Recent Cost Explorer Reports

- Amortized Costs

Cost Explorer Costs

This component tells you about two metrics: current cost comparison with the previous month's cost and current forecast comparison with the actuals of the previous month. The Cost Explorer shows the current cost of the month, until a certain date, as a chart and then compares that with the cost incurred for the same period in the previous month. It also forecasts for the remaining days of the month and then compares that with how much cost was actually incurred in the previous month. This helps the user decide on various factors, such as right-sizing, reducing the usage, etc.

Cost Explorer Trends

This section tells you about the cost trend of different services that a user is using. It shows the top trends in the dashboard, but the user can drill down to look at the trends of all the services and can further drill down to a particular service and look at the different costs that have accounted for that trend.

Daily Unblended Cost

First we must understand what an unblended cost is. It is the cost charged by AWS to a user, at a particular moment of time. For example, if AWS charged me $100 for using an EC2 instance at 10:45 a.m., then it will contribute to my Daily Unblended Cost. Generally, it is considered the most important cost data, as it tells you about the cost at the time it occurred. Obviously, you can use different filters to change the view of your Cost Explorer's unblended cost section. You can download the information as well in the form of a CSV. One thing should be clear: unblended cost does not show the refund amount that a user has received in a specific period.

Monthly Unblended Cost

As the name suggests, instead of looking at the Daily Unblended Cost information, you can change the granularity to a monthly level. This can give you an overall picture for the months, and you can then proceed to look at the Daily Unblended Cost of a specific month where you may find anomalies or you want to investigate further.

Net Unblended Cost

Net cost considers the total cost that has been incurred to you, adjusts for the discounts that have been given to you for the same period, and then presents you with the information. If a user wants, they can actually include or exclude any other adjustments such as refunds or credits given by AWS.

Recent Cost Explorer Reports

This metric includes a list of all the reports that a user has accessed in the past, with timestamps and links to each. You can just click the link and view the report again.

Amortized Costs

Generally amortized cost is a kind of normalized cost information. For example, if we consider a month, different service usage costs will be charged daily, and we can take a look at that in the unblended cost section. But, if we have taken a reserved instance, for example, then probably every first day of the month the entire month's cost will be charged. You will see a sudden spike there in the chart. Therefore, instead of dealing with all these sorts of spikes, the information can be normalized and distributed, and then it can give us a better overall picture. Figure 2-4 shows us the difference between unblended cost and amortized cost perfectly.

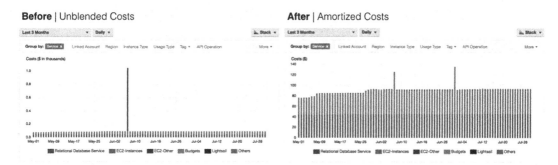

Figure 2-4. *Difference between unblended and amortized costs*

Now that we have looked at how AWS Cost Explorer works, let's explore how AWS Trusted Advisor works as well.

AWS Trusted Advisor

As we already know, AWS is the market leader when it comes to cloud computing services. Since it has a lot of clients, it knows the usage pattern statistics of everyone. Based on that, there are some best practices that AWS has come up with. They fall into the following categories:

- Cost optimization

- Security

- Fault tolerance

- Performance

- Service limits

We have already seen some of the factors that influence cost optimization in this chapter. We will be talking about the remaining factors in the next chapter. But, keeping all these factors in mind, a user can be charged a minimum cost with the maximum performance. AWS Trusted Advisor takes into consideration all these factors and provides its own recommendations that would help users to attain the minimum cost possible. How Trusted Advisor works is perfectly represented in Figure 2-5.

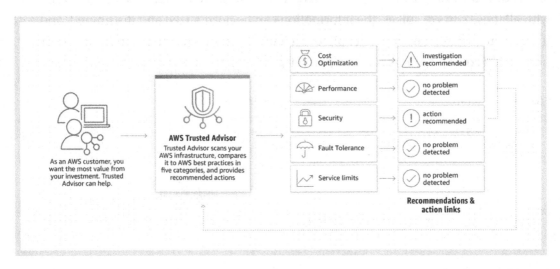

Figure 2-5. *AWS Trusted Advisor*

It works just like the antivirus protection software in your personal systems, which provide you with the malware and other related insights. Here, instead of malware insights, you get insights on the five categories that I specified earlier. The insights can look like Figure 2-6.

Figure 2-6. *AWS Trusted Advisor insights*

A green check mark specifies things we are doing in the right way. Yellow exclamation points are the warnings that need to be catered to or they may lead to problems. Red exclamation points mean that the recommendations should be looked at immediately as they are already affecting the cost, performance, security, etc.

The following are some of the benefits of using AWS Trusted Advisor:

- Trusted Advisor provides you with live notifications once you log in to your AWS account. If you enable it, then it can also send you periodic notifications via email.

- If you want, then you can make changes in the reports generated by Trusted Advisor. You can include or exclude items.

- When the Trusted Advisor makes a recommendation, it provides a link that takes you to the AWS Management Console where you can sort out the warnings.

Pricing of AWS Services

Now that we have looked at different ways of finding out the cost incurred for using AWS services and how to optimize them, let's look at where to find more information; see Table 2-1.

Table 2-1. *Cost Links*

AWS Service	AWS Pricing Page
On Demand EC2 Instances	https://aws.amazon.com/ec2/pricing/on-demand/
Spot EC2 Instances	https://aws.amazon.com/ec2/spot/pricing/
Reserved EC2 Instances	https://aws.amazon.com/ec2/pricing/reserved-instances/pricing/
Amazon S3	https://aws.amazon.com/s3/pricing/
Amazon VPC	https://aws.amazon.com/vpc/pricing/
Amazon SageMaker	https://aws.amazon.com/sagemaker/pricing/
Amazon RDS	https://aws.amazon.com/rds/pricing/

For information about other services, you can log in to the AWS Management Console and then switch to the service of your choice and look at its pricing section.

Conclusion

In this chapter, you learned how AWS charges for its services and how you can effectively use some of the AWS services to minimize your costs. This chapter helps you decide which services to opt for according to your budget and needs.

In the next chapter, you will be looking at the security aspects of AWS in detail. You will also look at how AWS handles fault tolerance and how you can effectively make an architecture that will serve the AWS services based on organizational hierarchy.

CHAPTER 3

Security in Amazon Web Services

In this chapter, we will look at the security aspects of AWS. As you know, AWS has hundreds of services serving thousands of customers, so a small compromise in its security could lead to a huge loss for a particular company and therefore even for AWS itself. That's why AWS has some concrete security practices, along with dedicated security services that take care of the entire umbrella of AWS features.

This chapter is dedicated to some of the most important parts of these services. By the end of this chapter, you will understand the underlying security of AWS and will also be able to implement different security features in the products you are using.

The SSR Model of AWS

When it comes to security, AWS follows the shared security responsibility (SSR) model. This model is simple: the responsibility should be shared between the customer and AWS. Specifically, AWS is responsible for the security of the entire infrastructure that it lends to its customers. The customer, on the other hand, is responsible for whatever it keeps in that infrastructure. Figure 3-1 shows the SSR model of AWS.

© Himanshu Singh 2021
H. Singh, *Practical Machine Learning with AWS*, https://doi.org/10.1007/978-1-4842-6222-1_3

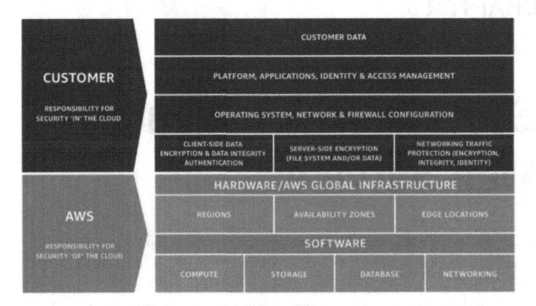

Figure 3-1. *Shared security model of AWS*

The following are some of the responsibilities of AWS:

- AWS takes care of the entire infrastructure, which includes the hardware, software, networking, and other facilities that AWS uses to run its services.

- AWS provides reports about its operations regularly by organizing third-party audits. Its compliances, security standards, and regulations are all verified.

- For most of the services, basic security tasks are also done by AWS, such as installing guest operating systems, patching databases, configuring firewalls, and doing disaster recovery.

The following are some of the responsibilities of customers and users:

- After the services have been booked and the basic security has been taken care of by AWS, it is the customer's responsibility to perform all the necessary security configuration and management tasks. This may include updating the OS, installing security patches, configuring the firewall, etc.

- The customer must set up user accounts using the Amazon Identity and Access Management (IAM) service. This will allow the customer to give each of its users different credentials and hence access to different services.

- The customer must set up multifactor authentication for each account.

We will discuss Amazon IAM and multifactor authentication in more depth later in this chapter.

AWS provides high-level security to its customers, which can be grouped into these categories:

- Compliance requirements

- Physical and environmental security

- Business continuity management

- Network security

Compliance

AWS's infrastructure is based on several IT security standards, including the following:

- SOC 1/SSAE 16/ISAE 3402 (formerly SAS 70)

- SOC 2

- SOC 3

- FISMA, DIACAP, and FedRAMP

- DOD CSM Levels 1–5

- PCI DSS Level 1

- ISO 9001/ISO 27001/ISO 27017/ISO 27018

In addition, AWS follows some industry-specific standards.

- Criminal Justice Information Services (CJIS)

- Cloud Security Alliance (CSA)

- Family Educational Rights and Privacy Act (FERPA)

- Health Insurance Portability and Accountability Act (HIPAA)

- Motion Picture Association of America (MPAA)

Physical and Environmental Security

For the physical and environment safety of the infrastructure, AWS follows these standards:

- Only authorized people are allowed inside the premises of AWS data centers, even if they are employees of AWS or Amazon.

- Automatic fire detection sensors and suppressants are installed in the facility.

- Not only do AWS data centers have a consistent electrical power supply, but they also have a powerful UPS so that in case of a power failure, the operations never stop.

- There are temperature control devices installed and continuously monitored by personnel so that the optimal temperature is maintained inside the data center.

- AWS decommissions any hardware following strict norms so that the data is never compromised.

Business Continuity Management

Lots of companies are moving to cloud platforms, and they do not their business operations interrupted because of AWS. When companies move their entire business infrastructure to AWS, all of their data resides there. Even a few hours of interruption can cause huge losses to companies. Therefore, AWS needs to be stringent in devising its infrastructure plan so that this does not happen. To ensure business continuity management, AWS has taken a lot of measures.

- All the data centers of AWS are live 24/7/365. So, if a particular data center fails, then the first thing AWS does is to move the data traffic away from the affected area.

- Using AWS, customers can put their data into multiple geographic regions, and inside each region the data can be put into multiple availability zones. The availability zones are made in such a way that they are independent from all the other availability zones in the same region. So, if one of the areas where the availability zone is present has a chance of being affected by floods in a specific season, for example, then all the remaining availability zones in the same region are put in locations that will never be affected by the same flood.

- Each data center is backed up by UPS and on-site backup generation facilities. Also, to further improve the security, power from different gridlines is provided inside the facility so that if one fails, the other remains effective.

Network Security

As we saw in the previous section, AWS has made its data centers resilient to calamities. But, a threat bigger than that is intrusion on the network. As technology advances, so do the skills of hackers. So, how does AWS tackle its network security? Well, let's explore some of the measures that AWS has taken to make its network secure.

- There is a firewall that secures your incoming and outgoing traffic. This is the most basic yet most powerful first step for network security.

- AWS has access control lists (ACLs) that contain a set of policies that decides who can access what services of AWS and what information goes to which service. These ACLs are regularly updated, and they are automatically pushed using the AWS tool called ACLManage.

- There can be a lot of entry points from where the information can flow into AWS servers, and the same is true with the exit points. Therefore, in such a scenario, managing those entry and exit points can become really difficult. Therefore, to deal with this issue, AWS has secure access points. Using them, AWS has only a limited number of access points through which comprehensive monitoring of all incoming and outgoing transmission takes place. These access points are called *API endpoints*, and only HTTPS access is allowed to have a secure connection.

- AWS provides the option of additional security to its customers through Amazon VPC. We discussed VPC in Chapter 1.

- For AWS there are two types of clients. One is its regular customers, while the other is its own corporate network. AWS doesn't give any special privileges to its corporate network while accessing the AWS services. For AWS employees, if they want to access any AWS service first, they have to raise a ticket, and if it gets approved, then they can get access. Also, for security purposes, everything that the employee does is logged securely.

- There are many automated monitoring tools that AWS has so that it can monitor the server and network usage, scanning the ports for inbound and outbound transmission, detecting unauthorized intrusion attempts, etc.

AWS Account Security Features

In the previous sections, we saw how AWS has made its infrastructure secure. In this section, we will look at how AWS makes its customers' AWS accounts secure. The first thing that AWS does is to secure every account with credentials. There are different ways in which AWS uses credentials for authentication.

- Passwords
- Cryptographic keys
- Digital signatures
- Certificates
- Multifactor authentication

Once you have created the credentials for your account, it is easy to download the report of credentials from the Security Credentials page. The information that is included in the report tell us about whether the account uses a password, whether the password is retirable, when was the password last changed, when the keys were last rotated, and whether multifactor authentication is enabled.

In AWS we can create multiple access keys and define multiple certificates as well. This is done to rotate them continuously for security reasons. When they are being rotated and we want business continuity to not be affected, we can use concurrent keys and certificates. We can use AWS IAM to rotate the keys. We will learn about the keys and certificates in more detail in the next section.

Passwords for Authentication

If you want to access AWS services, passwords are really important. They grant you the first level of access inside AWS. Passwords are created at the time of account creation and can be changed any time through the Security Credentials page. AWS allows passwords of up to 128 characters, and they must have special character combinations to be strong enough.

If your organization's infrastructure is entirely hosted on AWS, you can create password policies so that new passwords must follow your security policy. This assures that the strongest of passwords are created.

Multifactor Authentication

This is an additional level of security that a customer may opt for. Multifactor authentication requires that after the customer has successfully entered the username and password, the user will have to provide a specific and unique six-digit code for authentication. If that is successful, then the person will be allowed to enter the account. This six-digit code is received by the customer to one of its authenticated devices. It can be a smartphone, email, or phone number. As the person logs in, a code is received that should be entered, and then successful login takes place.

Access Keys for API Authentication

An API is a feature where a request is sent to a piece of code encapsulated behind the API and the person receives the output. Now, the coding logic inside an API may be allowing the user to access sensitive information. So, the request that is sent to these APIs must be authenticated as well, which means only the right user must get the access to the API. This is done through a *digital signature*. This digital signature is generated by passing the request text and secret access key to a hash function responsible for the cryptography. This hash function encrypts the message and then sends it to the API. This in turn gets decrypted at the API end; the secret key is checked, and then the entry is provided.

Currently, digital signatures use a protocol named HMAC-SHA256 that is in its fourth version. One more level of security in digital signature verification is that a timestamp is added to the request. If the timestamps of the digital signature being generated and being received by the API are different by greater than 15 minutes, the request is denied.

X.509 Certificates

When we want two or more web services to talk to each other, we use SOAP-based requests. To make these sorts of requests secure, we use X.509 certificates. These certificates consist of three parts.

- A public key

- A private key

- Additional metadata

The first step is to generate the digital signature by using the process that we looked at in the previous section. Now, we use this digital signature and the certificate to send the request. First, AWS tries to verify that the authenticated user is sending the request by decrypting the digital signature and verifying it. After that, the certificate that has been sent is matched by AWS with the certificate uploaded by the authenticated user in their own AWS account. If everything is green, the request is sent forward; otherwise, it's denied.

AWS Identity and Access Management

Just like we have the Amazon Management Console to access all the AWS services, similarly for all the security requirements we have AWS Identity and Access Management. All authentication or authorization can be managed from IAM. We can make users, define roles, give permissions, assign policies, and do a lot of other things, all from IAM. Hence, this is a service that everyone who is using AWS, either for machine learning or for web development, must know about.

The first thing that a person looking to use AWS services has to do is to create a root user account. This account can be accessed through a username and password, and it gives the users indefinite access to all the services that it has been registered to. But, AWS discourages us from always logging in using the root user account as it can compromise the security of the organization. That's why, instead of using the root user, we can create different users using the IAM service. These are called *IAM users*. The admin can give permissions to all these users, ask them for their own passwords, assign different policies, and hence make the entire infrastructure simple yet secure.

To understand this in a much better way, let's look at an example. Suppose there is an organization with 5,000 employees. There will be a board of directors, CEO, CTO, presidents, managers, architects, engineers, and other employees. First, we cannot give everyone root access, which should not require any explanation. So, the organization can create IAM users. But, the permissions given to the top-level management will be different from the permissions to the employees lower in the hierarchy. Even in the top management, maybe the people on the board of directors have only read-only access to specific services that have dashboards or visualization supports (for example Splunk Dashboard), but a CTO has a kind of root access. Similarly, the engineers will have access to the services that they have been hired for. Machine learning engineers may have access to SageMaker, data engineers may have access to EMR or DynamoDB, and so on. Hence, we can conclude with this example that different users require different permissions, and all of this can be done using AWS IAM.

Once an IAM user is defined, it will have its own password. Once the person logs in, the user will see only those services they have permission to use. Other services will ask for credentials or deny the access. Note that an IAM user may not be an actual person. It could be a software service. For example, say a company has made a website hosted on a different platform, but uses certain AWS services like S3 or DynamoDB. Hence, the website should continuously be talking to these services. To authenticate that the right software talks to the AWS services, an IAM user is made for these services, and the website accesses the respective services with its own credentials.

But, organizations do already have their own infrastructure, and each employee from any hierarchy of management has an email ID and password. When the employee has to use AWS, they will have a new ID and password. That means the person will have to remember and secure two accounts: the corporate account and the AWS account. Wouldn't it be awesome if AWS provided a way so that an employee just needs to log in to the corporate account and automatically get logged in to the AWS account? AWS does have a solution for this called *federated users*.

Federation of Users in AWS

If an employee has an account in the corporate network and a corporation has an account in the AWS network, then once the employee logs in to the corporate network, the user identity can be federated by the organization to the AWS network. This way, it eases the login hassles for corporate employees inside AWS. See Figure 3-2.

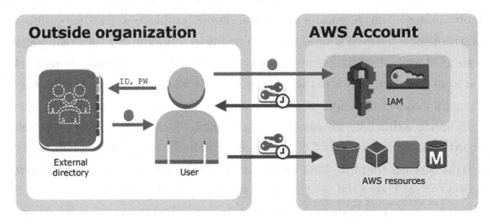

Figure 3-2. *Federation of users*

The following are some features of federation:

- Corporate networks use *single sign-on* (SSO) that makes the federation possible. After this, the automated login happens, and the employee reaches the Amazon Management Console without having to log in again.

- Even if a person has accounts in platforms such as Gmail, Facebook, Amazon, etc., they can be used for SSO.

How Access Management Is Done in AWS

In common terminology, *access management* can also be called *authorization*. A person or a service that logs in to the AWS account is called a *principal entity*. The principal entity is authenticated using IAM (it is used to authenticate an IAM role or a user). Once the authentication is done, the policies attached to the particular user or role are checked. These policies help determine what permissions are given to the principal entity. They are created and attached to different IAM identities such as users, roles, or groups. Now, if the principal entity sends a request for a service but the permission for the usage of that service is not defined in the policies, then the access will be denied. This is how access management is done in AWS.

Policies can be applied based on IAM roles, resources, or access control lists. An example of a role is a data engineer. All the IAM users with the role of data engineer will be assigned the same policy. This means everyone in that role will be able to access the same resources with the same level of permissions. Policies that are applied on resources put restrictions on the usage of a specific resource. For example, we can create a policy that gives users the permission to read a DynamoDB table but not create one. ACLs are used when we want to apply cross-platform policies.

An example of a role-based policy is given here. We have defined a JSON file that shows the policy defined on DynamoDB.

```
{
  "Version": "2012-10-17",
  "Statement": {
    "Effect": "Allow",
    "Action": "dynamodb:*",
    "Resource": "arn:aws:dynamodb:us-east-2:123456789012:table/Books"
  }
}
```

This gives the user permission to use the Books table present in DynamoDB. This makes it clear that to assign policies, JSON files must be created, and these files must be attached to respective entities. Policies not only can be attached to specific entities, but to a group of entities as well. These are called *IAM groups*; see Figure 3-3.

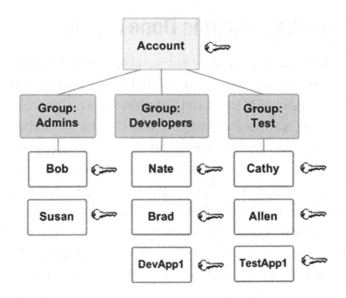

Figure 3-3. *IAM groups*

Figure 3-3 defines multiple IAM users and groups them into three IAM groups: Admins, Developers, and Test. Now, the policies can be applied on the entire group, which gets automatically applied to its members. But, one thing to notice here is that if a specific member has not been given permission for a service, then even though the group has access to that service, the member will still not have permission to access that service.

Where can we find a summary of all the policies defined? There are three tables in the IAM console that give us detailed information about policies. These tables are as follows:

- Policy summary

- Service summary

- Action summary

When we open the policy summary, we will see a list of services on which the policies have been defined. We can click any of the services to go to that particular service's summary table. The summary table tells all the actions that can be performed on that particular service and the permissions attached to those actions. You can click any of the actions, and then you can come to the action summary that gives the permissions that have been granted to that particular action. Figure 3-4 sums these tables up for you.

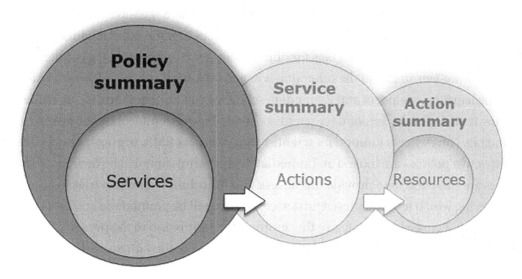

Figure 3-4. *Policy, service, and action summaries*

Policies can be divided into two kinds: identity- and resource-based policies. As the names suggest, when the policies are attached to IAM roles, users, or groups, they are called *identity-based policies*, whereas when we attach them to resources such as S3 buckets or DynamoDB, they are called *resource-based policies*.

Attribute-Based Access Control

Before talking about attribute-based access control (ABAC), we must understand *AWS tags*. These are some attributes that are attached to IAM users, roles, groups, or even AWS resources. These tags are useful when we have a big organization and we want to make policies for the employees. How are tags useful? Let's look at an example.

A company making machine learning products has shifted its entire architecture to AWS. Now it is making models on Amazon SageMaker, doing data engineering using Amazon EMR, making websites, APIs, and doing database management using DynamoDB and CloudFormation, and for security using various other AWS services such as CloudWatch, TrustedAdvisor, etc. Now, let's also assume that the company has about 5,000 employees. Providing IAM roles and then attaching policies to every employee would be a tedious task. So, what do we do? That's when tags and ABAC come into the picture.

Imagine two machine learning engineers; one only uses SageMaker and S3 buckets, while the other one uses EMR in addition to the services that the other engineer uses. The organization can create two tags for different usages. Suppose one tag is called *basic-machine-learning*, and the other one is called *advanced-machine-learning*. All the machine learning engineers and junior data scientists can be given the first tag while defining their IAM role. Senior data scientists and solution architects can be given the second tag. Now when a junior data scientist tries to access EMR, the tag is checked, and its respective policies are looked at. Immediately the permission will be denied, as the policies are not enough to provide this access. But if a solution architect tries to access the same tag which junior data scientist accessed, he will be granted the access as the associated policies with the tag give the appropriate permission to the person. These tags are also called as *attributes*, and hence this process is called *attribute-based access control* (ABAC).

How can we say that the ABAC process is better than the traditional process? Let's look at some of the differences between them. The traditional process is also called *role-based access control* (RBAC).

- The first difference is that in RBAC every time a new resource is added, the policies attached to a role or user must be updated—not only for one role, but for all the roles. Instead, in ABAC, only the tag needs to be updated.

- Because of these tags, ABAC has much fewer policies compared to RBAC and hence is easier to manage.

In the second part of this book, we will look at the entire process of creating the root credentials and then defining roles, users, groups, etc., using IAM. In this section, we have covered the theoretical aspects of it. Practically we will be looking at the applications in the next section.

There are some other services that AWS provides to increase the security of the applications.

- AWS WAF

- AWS Shield

- AWS Firewall Manager

Let's discuss each service one by one.

AWS Web Application Firewall

All the HTTP or HTTPS requests that are sent to specific AWS services like Amazon API Gateway, Amazon CloudFront, or Application Load Balancer are monitored by AWS Web Application Firewall (WAF). WAF allows you to control the access to the content. It does this based on certain conditions and rules. Some of these rules are listed here:

- Allow every HTTP/HTTPS request except the ones that are explicitly specified.

- Block every HTTP/HTTPS request except the ones that are explicitly specified.

- Count the requests and match the properties defined in the requests with the properties mentioned in WAF. If the count of the properties that are the same matches, then allow the requests; otherwise, block them.

The conditions that are matched to allow or block the requests follow some characteristics. The following are some of those characteristics:

- Monitor the IP address to see where the request originated.

- Look at the country from where the request originated.

- Analyze the values present in the request headers.

- Regular expressions can be made that search for specific patterns in the string and then make the decision of acceptance or rejection.

- Check whether SQL code is present, which can be malicious.

- Check whether scripts are present, which can be malicious.

WAF can also be used to protect the applications that are hosted inside ECR. ECR allows you to efficiently manage Docker containers inside clusters. To define the rules and conditions in WAF, the following features can be used:

- Web ACLs

- Rules

- Rule groups

First, an ACL can be created that monitors the access to specific AWS resources. Then rules can be assigned to these ACLs that will act as a firewall and monitor the requests. Each rule is a kind of a statement with a condition, which is termed the *inspection criteria*. If the condition is met, then either the requests are allowed or they are blocked. We can also use rule groups, which are groups of statements containing conditions that can be attached to your ACL and hence indirectly to AWS resources.

AWS Shield

Before looking at the AWS Shield service, we must first understand a cyberattack called a *distributed denial of service* (DDoS), as AWS Shield helps to mitigate it.

Every web service or server has some bandwidth to serve as much of its current user base as possible, while handling more traffic during peak times. To conduct a DDoS attack, cybercriminals flood the service or server with so much traffic that either it becomes difficult for the users to operate or the entire service or server crashes. If the traffic is generated from a single system, the attack will not be that effective. That's why the traffic is generated from multiple systems in parallel, and the target is attacked. Figure 3-5 shows a visual representation of a DDoS attack.

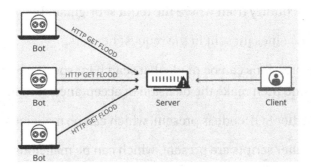

Figure 3-5. *DDoS attack*

AWS WAF can be used to block these sorts of attacks, but for advanced security we can also use AWS Shield Standard and AWS Shield Advanced. By default all the services that we use in AWS come with AWS Standard Shield, at no extra cost. To use AWS Advanced Shield, users need to pay a little extra.

There are two specific layers in a network that get compromised with cyberattacks: the network and transport layers. AWS Standard Shield monitors both these layers and provides protection. If AWS Standard Shield is used with Amazon CloudFront or Route53, then it provides some additional benefits as well.

For additional protection, for example, if you want to protect applications running in EC2 or Elastic Load Balancer, a user can opt for AWS Advanced Shield. Advanced Shield provides security not only for the network and transport layers but also for the application layer. Figure 3-6 shows the different layers inside a network, using the Open Source Interconnection (OSI) model of networking.

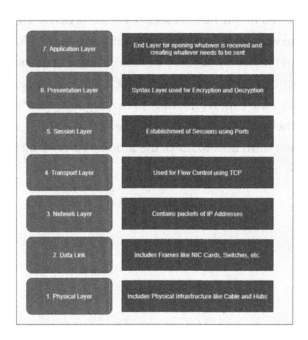

Figure 3-6. *OSI seven-layer model*

AWS Firewall Manager

To ensure AWS WAF and AWS Shield operate smoothly and efficiently, AWS Firewall Manager can be used. AWS Firewall Manager is responsible for automatically applying all the rules defined on the resources and services. This holds true even when new resources are added in AWS. Because of this automatic capability, AWS Firewall Manager provides lots of advantages.

- If a group of resources is following a specific tag, AWS Firewall Manager can automatically apply some custom rules.

- AWS Firewall Manager allows you to create and manage your own rules or the rules bought from the marketplace.

- It is generally beneficial to use AWS Firewall Manager if you have many users in an organization.

Conclusion

In this chapter, you learned about various security aspects of AWS. This finishes our first part of this book, where we covered all the basics related to AWS. Now, we will move on to the second part of the book where we will look at how to make machine learning models using AWS. In the next chapter, we will look at the concepts from this chapter more practically and become experts not only in making the machine learning models but in using various other services such as S3 buckets, DynamoDB, the security tools of AWS, etc. You will also learn how to make the models efficient and how they can be automated.

PART II

Machine Learning in AWS

CHAPTER 4

Introduction to Machine Learning

This chapter covers two main topics. First, you'll be introduced to machine learning and its components, and then, you'll look at the different services that AWS provides to help you make machine learning models.

Introduction to Machine Learning and Artificial Intelligence

Machine learning and *artificial intelligence* are two terms that are used a lot in the industry nowadays. Most of the time people think they are synonyms of each other with no major differences. However, they are different. *Artificial intelligence* is the field of computer science and mathematics that tries to mimic human-like behavior and make decisions similar to humans. *Machine learning* can be considered a subset of artificial intelligence where we make a machine learn through the historical data provided and then use the learned behavior to predict an outcome if we get similar information in the future. So, machine learning is all about prediction and recommendations, while anything where we enable a machine to think like a human is artificial intelligence.

For example, we converse with Alexa just like it is a human. The Alpha Zero robot can defeat chess champions as if it was a super-intelligent human being. Cars can now drive automatically without the help of humans, making their own decisions live on the road. All these are the examples of artificial intelligence agents, because they are behaving just like humans.

When we try to forecast sales for the next six months, predict whether a stock is going to move upward or downward in the coming days, or guess whether a machine is able to understand the context of a paragraph and classify whether it is talking negatively or

© Himanshu Singh 2021
H. Singh, *Practical Machine Learning with AWS*, https://doi.org/10.1007/978-1-4842-6222-1_4

positively about certain things, we are talking about machine learning. Here we have trained the models by giving past data, and then the machine predicts the outcome of the future incoming data.

Why exactly do we want machines to learn and assist us in our day-to-day lives? The following are some of the compelling reasons:

- We as humans make decisions by learning from our surroundings. But we are limited to this exposure to information. Also, the chances of making mistakes as a human for new scenarios are high. That's why we have the saying, "After all, I am only human." Machines, on the other hand, can be given as much information as possible. They can be given as much complex information as possible. Hence, after getting trained, they can make their own decisions based on the surroundings. Of course, we cannot have a machine make decisions independently, but they can assist humans in decision-making.

- In the past, people were making systems that were rule-based. This means if a certain scenario occurs, then do this; otherwise, do that. But, today, data is becoming complex, and all four Vs of data—volume, veracity, velocity, and variety—are at their peak. Therefore, making rule-based systems is next to impossible. Machine learning systems help us to understand this complex unstructured data and make decisions.

Machine learning solves these problems with three types of learning.

- Supervised learning
- Unsupervised learning
- Reinforcement learning

Supervised Learning

Supervised learning is a branch of machine learning where we know exactly what has happened in the past, and then we try to predict if the same result will occur if some mix of the situation that happened in the past occurs again. For example, for the past six years we have collected the rain and snow data for some villages. Specifically, we collected different information about humidity, pressure, temperature, etc., whenever it rained or snowed. We trained a machine learning model based on that data. Now, if we know the humidity, pressure, and temperature information for the upcoming days, can we predict that it's going to rain or snow? This is supervised learning.

In supervised learning terminology, the factors of humidity, pressure, temperature, etc., are called the *independent variables*. The factor that we are trying to predict, that is, whether it is going to rain or snow, is called the *dependent variable*.

Figure 4-1 shows a supervised learning framework. You can see that there is a supervisor who checks whether the predicted output is the same as the expected output. Based on that, an adjustment is made, and this adjustment is different for different machine learning algorithms. Finally, we get our predicted output.

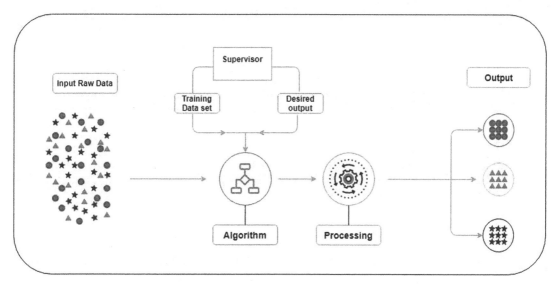

Figure 4-1. *Supervised learning*

Unsupervised Learning

In this branch of machine learning, we don't have a guiding variable, which means that the dependent variable is not present. Hence, the main aim of unsupervised learning is to first understand the patterns present inside the data and then put the data that follows a similar pattern in the same group. Therefore, we can have clusters with similar features, or we can have similar products that are bought together, etc. Let's understand this with the help of an example.

When we go to a retail shop, such as Ikea, we can see a lot of products. We have gone there with the intention of buying a specific item. We also buy some other items. At the time of billing, the cashier will ask us for our mobile number through which they can identify unique customers. Now, since we liked the product, we went to the shop again and bought other necessary things. This shop has become our favorite one, and all our shopping is done there for three years. Using unsupervised machine learning, now the shop can utilize our purchase history and can start recommending different products. For example, if we recently bought curtains, it may suggest some other decorative items.

Figure 4-2 shows the unsupervised learning framework. In this diagram you can see that in the Interpretation section a hidden pattern of similarity is found, and based on that, similar items are put in the same place, as we can see in the output.

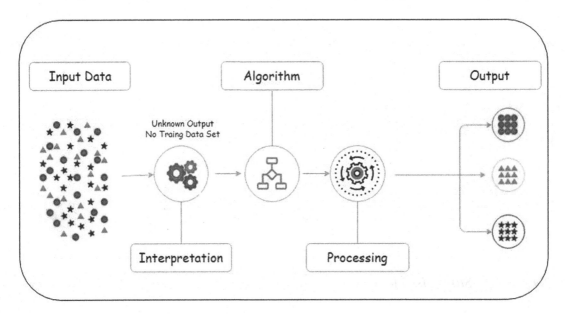

Figure 4-2. *Unsupervised learning*

You can find this sort of example not only in physical shops but also on e-commerce websites.

Reinforcement Learning

Reinforcement learning is a unique domain of machine learning where a machine tries to learn by itself by analyzing different scenarios. This is done by giving the machine rewards for every successful task it does and giving a penalty if it fails to do a task. The machine has the aim of getting as many rewards as possible. With this model, machines learn to operate in different scenarios automatically. Let's look at an example.

A robot wants to learn to walk. First the environment provided to it is a plain road. For every successful step, maybe 50 reward points are given, but for every fall, 100 penalty points are taken. The robot takes the first step and falls, so it receives the penalty. It does this multiple times and keeps on falling. Finally, it takes a step and doesn't fall. Hence, it receives reward points. This time, the robot knows what it has done to get its first reward. Now, it will try not to fall every time so that it can get the maximum number of rewards. Once it has trained itself to learn to walk, the environment can be changed, and multiple obstacles can be introduced. Again, the robot will start exploring different options to get the maximum rewards. This is how reinforcement learning takes place.

In reinforcement learning, there is a combination of two approaches: exploration and exploitation. *Exploration* means that the robot should look at multiple options in the environment so that it can learn every difficulty present. *Exploitation* means that whatever the robot does, it has to keep getting the maximum number of reward points. The robot explores and maximizes the reward points, and hence it learns how to operate in tough situations. Reinforcement learning is out of the scope of this book, but we will be covering other aspects of machine learning in detail in this book.

Figure 4-3 shows what a reinforcement learning process looks like.

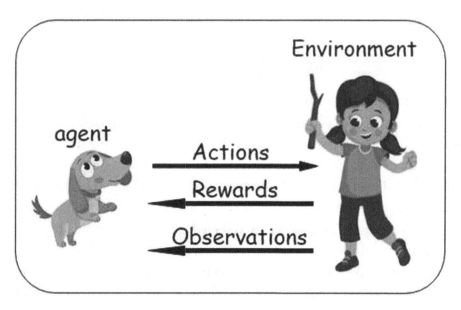

Figure 4-3. *Reinforcement learning*

There is one more type of machine learning that requires a dedicated discussion: deep learning.

Deep Learning

What do we humans exactly think? How do we make decisions? Obviously, it is the brain that is responsible, but it is not entirely responsible for making decisions. The brain will only perform when some information reaches it, but to make this information reach the brain, there are some other very small yet important components responsible. These are called *neurons*. They are the first ones that receive the information, and then through a series of neurons, this information is transferred to the brain, which in return makes a decision. This chain of neurons is called a *neural network* or *biological neural network*. Figure 4-4 shows a simple biological neuron and its components.

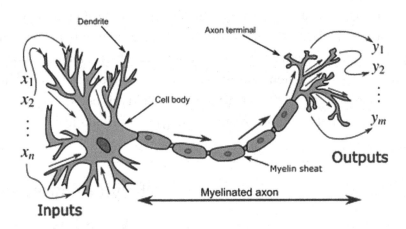

Figure 4-4. *Biological neuron*

Mathematical and computer experts researched whether it is possible to mimic this human behavior of receiving information and then making decisions. This research led to the field artificial neural networks, which constitutes a major part of deep learning. These neural networks perform in a similar way to how a biological neural network performs. They receive the information, and then through a series of mathematical equations, like forward and backward propagation, gradient descent, activation functions, etc., they make the decisions. We will be discussing this later in the book.

Figure 4-5 shows an artificial neural network consisting of multiple artificial neurons present inside hidden layers.

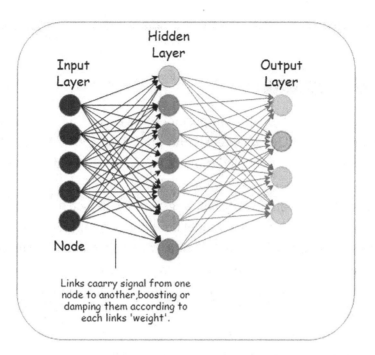

Input Layer

Hidden Layer

Output Layer

Node

Links caarry signal from one node to another,boosting or damping them according to each links 'weight'.

Figure 4-5. *Artificial neural networks*

Now that we have seen a general introduction to machine learning, let's explore the different services of AWS that help users to build machine learning models.

Machine Learning in AWS

We will be discussing the following AWS services in this section:

- Amazon SageMaker

- Amazon Comprehend

- Amazon Polly

- Amazon Rekognition

- Amazon Ground Truth

- Amazon Textract

- Amazon Translate

- Amazon Transcribe

- Amazon Lex

Let's start by exploring the first, and the most important service, Amazon SageMaker.

Amazon SageMaker

Amazon SageMaker is one of the most important services used across industries. Therefore, it is the base of all the chapters in this book, and understanding the service is imperative, so I will be giving a detailed explanation of it compared to the other machine learning services of Amazon.

Machine learning is not about only building a model; in fact, in my experience, a minimum amount of time is given to model building as compared to feature engineering, data preparation, or model serving. SageMaker makes the life of a data scientist much easier by providing services that you can use to prepare data, build models, test them, and then deploy them into production environments. It provides most of the common algorithms for building your machine learning models, and if you want to make any custom model not supported by SageMaker, then it has a facility to do so by using a bring-your-own container service. It also provides a distributed training option that can make your models run faster, as compared to a single-node run.

Amazon SageMaker comes with following features:

- **SageMaker Studio**

  This is an application where you can build, train, validate, process, and deploy the models. It's a single place to do everything.

- **SageMaker Ground Truth**

  This is used to create a labeled dataset.

- **Studio Notebooks**

  This is one of the latest features of SageMaker that includes the single sign-on feature, faster startup time, and one-click file sharing.

- **Preprocessing**

  This is used for analyzing and exploring data. It does feature engineering and transformation of data, as well as all the other things required to prepare the data for machine learning.

- **Debugger**

  This has different debugging usages, such as tracking the hyperparameters whose values keep changing during the model training. It can even alert if something abnormal happens with the parameters or with the data.

- **Auto-pilot**

  Without writing a single line of code, if you want SageMaker to take care of your model building, either regression or classification problems, auto-pilot is the feature to use. It is generally for users who have less coding experience.

- **Reinforcement Learning**

  This provides an interface to run a reinforcement learning algorithm, which runs on a reward and penalty architecture.

- **Batch Transform**

  After building the model, if you want to get predictions on a subset of data or you want to preprocess a subset of data, you can use the batch transform feature of SageMaker.

- **Model Monitor**

  This is used to check whether the model quality is persistent or deviates from the standard model.

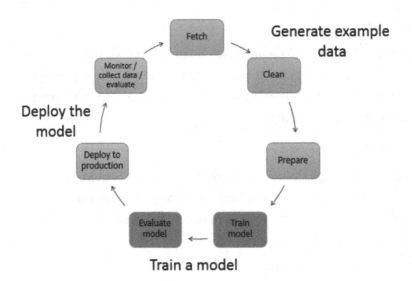

Figure 4-6. *SageMaker process*

Understanding How SageMaker Works

Figure 4-6 shows the stepwise process of how SageMaker works.

This diagram is valid not only for SageMaker, but for any machine learning models that we make. They all undergo the same process. These are the main steps that the process follows:

1. **Fetch data**

   This is the first step for building any machine learning model. Once we have decided on the problem statement that we have to solve, we have to accumulate all the data related to it. The data can be in the format of a database table, Excel sheets, text files, Word documents, images, etc. Once we know about all the data sources, these files need to be put inside a single repository so that the model knows about the location.

2. **Clean the data**

   Our data can have null values, outliers, misspelled words, corrupted files, etc. All these things need to be explored and sorted out before the data is being given to the model. There are a lot of statistical methods as well that are used for data cleaning,

3. **Prepare data**

   Once we have made our data clean, it is time to prepare our data. This includes all the transformations done on the data, scaling and normalization processes, combination of features or splitting of features, etc. After all these things are done, it has to be stored at a specific place so that the model knows the reference to the clean and prepared data files.

   The first three steps that we have seen, all these things can be done inside the SageMaker Jupyter Notebook, and after that, the cleaned data can be stored inside an S3 bucket.

4. **Train the model**

   Once the data is prepared, we need to train the model. The first thing is to select the model that needs to be applied. The models can be chosen from the list of built-in algorithms that SageMaker provides, or custom models can also be used by making your own containers and uploading them to AWS or buying them from the AWS marketplace.

   Also, for training the model, we must decide on what kind of computation is required. Selection can be made based on the RAM size or number of GPU counts, etc. It is decided based on how big the dataset is or how complex the model is.

5. **Evaluate the trained model**

   Once the model is successfully trained on the dataset, it needs to be evaluated before deploying it for production. For this, multiple metrics can be used. For regression models, RMSE scores can be used, while for classification models precision and recall can be used. Once the metric crosses the decided threshold, only then can it be moved toward production.

6. **Deploy the model to production**

It is easy to deploy the model in SageMaker. Generally, in normal
scenarios one has to make APIs and then serve the model through
an endpoint. For all this, coding requirements are necessary.
But, in SageMaker, with minimal coding efforts the model can be
converted into an API endpoint, and after that live or batch model
inference can be started. Also, to deploy the model, another
computational instance can be chosen, which generally takes less
RAM or GPUs as compared to the training model instance.

7. **Monitor the model**

Once the model starts serving in production, we can keep
monitoring the model's performance. We can measure for which
data points the model is performing well, as well as the areas it is
not. This process is called knowing the *ground truth*.

8. **Repeat the process when more data comes (retraining)**

Finally, as and when new data comes, the model can be retrained,
and all the previous steps can be repeated. All this can be done
with zero downtime. This means that the old model keeps serving
until the new model is put into production.

Preprocessing of Data in SageMaker

As we talked about in the previous section, before we give the data to any model, we first
clean it and preprocess it. We can do this in SageMaker in multiple ways.

- Using SageMaker Jupyter Notebook to write Python scripts for
  processing data

- Using a SageMaker batch transform script to process data before
  getting the inference

- Using Script Processor to write processing script on the data

Using one of these approaches, the data can be processed, and then any of the
SageMaker training models can be called to do the training on this processed data. We
can use popular Python libraries such as Scikit-Learn or TensorFlow for this purpose. If

your script involves some other libraries, then you can upload your own script inside a Docker container. You'll learn more about this in later chapters.

Model Training in SageMaker

Figure 4-7 shows how exactly model training happens as well as how the model deployment happens. In this section, we will talk about the training part, while in the next section we will cover the deployment part.

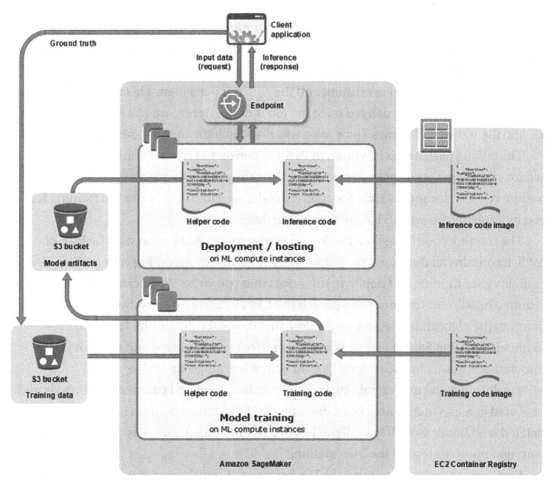

Figure 4-7. *SageMaker training and deployment process*

To understand how model training in SageMaker works, we will look at the bottom part of the image. We can see that there are five sections contributing to it.

- S3 bucket for training data

- Helper code

- Training code

- Training code image

- S3 bucket for model artifacts

Training a model in SageMaker is called a *training job*. Any algorithm that is executed in SageMaker requires the training data to be present in an S3 bucket. This is because the compute instances that are used for training the model are called dynamically during model execution, and they are not persistent. This means the data that is stored there will be deleted once the job is done. Hence, we can save the data in S3, and the model will always know from where to fetch the data, by means of an S3 URL.

The coding part, which is written in Python, consists of two sections. The first section, the helper code, helps you in processing the data, fetching the data, storing the output, etc. The second section, the training code, actually does the model training for you by applying the selected algorithm on the data.

The training code image is a Docker container image that is stored in the ECR of AWS. It contains all the packages and software required for executing your code. It also contains your training and deployment scripts that you write. We package everything required inside one container and push it to ECR. Then, we just pass the URL of the image to the algorithm selected, and automatically the training script runs. We need to understand that SageMaker works based on Docker containers, and hence it is imperative for users to understand Docker before learning SageMaker.

Finally, once the model training is done, the model-related parameter values should be stored in S3; as mentioned, once the training job is done, compute instances are deleted, and hence we will lose all our learned parameters. That's why S3 becomes the common point to store all the information.

One thing to notice here is that the Docker image is built by you, but still we have not selected the hardware requirements. Therefore, when we call the SageMaker algorithm and when we pass the parameters such as the S3 URL and Docker Image URL, then only can we pass the type of instance that we have to choose. These instances are the EC2 instances that we saw in Chapter 1. Once we have chosen the instance, the Docker image is downloaded on that instance, along with the training data. Finally, the model training starts.

We will look at all these aspects of training the model in SageMaker in the upcoming chapters.

Model Deployment in SageMaker

Once the model training is done, all the learned parameters are stored in the S3 bucket and called *model artifacts*. These model artifacts will be used during inference (or predictions). In Figure 4-7 the bottom part was the model training part; now we will discuss the upper part, which is the model deployment part. It consists of the following sections:

- URL reference to model artifacts in S3 bucket

- Helper and inference code

- Inference code image

- Endpoint

- Client

The helper and inference code consists of processing scripts and prediction scripts. Also, it includes the format in which the predictions need to be sent or saved. For the predictions, the model artifacts generated during the training part are used.

SageMaker removes the training compute requirements with the deployment compute requirements. This is because training may require big instances with stronger computational power, but for predictions we do not require that many big instances. Hence, the predictions can be done with smaller instances as well. This helps save a lot of cost.

We can use the same Docker image that we built for training a model for the inference by just adding a few extra Python scripts that help in deployment. That may include using packages such as Flask, Gunicorn, etc. To start the deployment, we need to pass the model artifacts the URL, the ECR image URL, and the compute instance that we need. By giving these three parameters, the deployment is made, and an endpoint is created.

The endpoint is a place where we send requests in a particular format, maybe CSV or JSON, and get the response from the model. This is called a RESTful API. The model that is created is served through this API, and the data on which we want predictions is sent as a CSV, row by row, and we get the predictions in the same way. These are POST and GET requests. We can expose this endpoint to any client objects. It can be a website, a mobile app, an IOT device, or anything else. We just need some records sent to the endpoint and to get the predictions.

Endpoints are used when we make live predictions. Hence, they keep running until and unless we manually stop them or add a timeout condition. But suppose we want the predictions for a subset of data, maybe 5,000 rows, and we don't want a live endpoint. Then SageMaker supports something called a *batch transform*. Using this approach, we provide the same parameters that we provided to deployment code, but one extra parameter is provided. It is the link to the data on which inference is needed. This data is again stored in S3 and hence downloaded to the instance when prediction is required. After the prediction is done, predictions are stored in S3, and then the compute instance is stopped immediately. Figure 4-8 shows the process of batch transform in SageMaker.

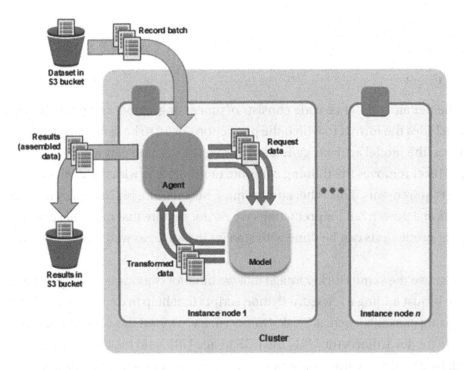

Figure 4-8. *Batch processing*

We will look at both the approaches, endpoint generation and batch transform, in the upcoming chapters.

Built-in SageMaker Algorithms

The following are all the algorithms that SageMaker provides:

- **Blazing text**

  This algorithm is used for problem statements involving text classification. It is an optimized version of the Word2Vec algorithm and can be used for multiple tasks such as sentiment analysis, named entity recognition, etc.

- **DeepAR forecasting**

  This algorithm is used in the domain of univariate time-series forecasting using RNNs. It can be used to train on multiple similar time-series data, and it outperforms the ARIMA or exponential smoothing methods in most of the scenarios.

- **Factorization machines**

  This is a general-purpose algorithm that can be used both for regression and for classification tasks. For classification, it only supports binary classification problems.

- **Image classification**

  This is built on ResNet (CNN model) for multilabel classification of images. It can be trained from scratch if the dataset available is big; otherwise, transfer learning can be applied if the size of the dataset is small.

- **IP insights**

  This algorithm is used for a special use case—finding the usage patterns for IPv4 addresses. It can be used to find out whether the IP address from which a user is sending a request is anomalous.

- **K-means**

  This algorithm is used to find clusters of data that are following similar patterns. It is an optimized version of the statistical k-means clustering algorithm.

- **K-nearest neighbor**

  It is used for the classification of data by using an approach finding the nearest neighbors. It is an optimized version of the k-nearest neighbor statistical algorithm.

- **Latent Dirichlet allocation**

  This is an algorithm used to find out topics present inside documents, and hence the domain of application is also called *topic modeling*. It is an unsupervised learning approach used to find out categories from a bunch of documents.

- **Linear learner**

  The is a normal simple and multiple linear regression algorithm having the capacity to perform logistic regression as well on the classification problems.

- **Neural topic model**

  This is again a topic modeling approach, where the topics are extracted from a bunch of documents by finding out their statistical distributions. This algorithm can be used in the domain on text summarization or recommendations.

- **Object2Vec**

  This is used for generating the vectors for the objects, and it is similar to the Word2Vec algorithm. The only thing is that it is a generalized version of it. Using this approach, a lot of optimized and efficient classification and regression can be made giving us better performance.

- **Object detection**

  Finding and recognizing objects present in an image is the main task of this algorithm. It contains a single deep neural network to perform this operation. The framework used in this model is single-shot multibox detector (SSD) and uses VGGNet and ResNet as a base.

- **Principal component analysis**

  This is based on finding the importance of variables and then combining variables based on the similarity. It is used for the dimensionality reduction of data so that the number of variables can be reduced based on combining their importance using the concept of eigenvalues and eigenvectors.

- **Random cut forest**

  This algorithm is used to find patterns present inside the datasets and then find those patterns that deviate from all the general patterns present in the dataset. For example, why is there an unnecessary spike in time-series data? Why is a particular data point not being able to be classified? These are a few of the multiple uses of the random cut forest method.

- **Semantic segmentation**

  This is used for developing computer vision applications. It is a pixel-level approach algorithm where each pixel is labeled based on the data. It can be used in the domain of self-driving cars, medical imaging, etc.

- **SeqtoSeq modeling**

  This is used when we have a sequence of input and we have to generate a sequence of outputs. For example, we can have sequences of input in German and need to translate the sequence into English. Similarly, it can be used for time-series data, images, and other text applications. This algorithm uses RNNs and CNNs with an attention-based approach.

- **XGBoost**

  This is one of the highly optimized versions of ensemble trees that uses the concept of gradient boosting approach as well as takes the power of multithreading and multiprocessing to give awesome inference on the datasets. It is one of the most used algorithms, not only in SageMaker but elsewhere.

We will be seeing most of these algorithms and their practical implementations in the upcoming chapters.

Custom Algorithms in SageMaker

In the previous section, we saw the different algorithms that SageMaker supports. But what if you want to use a different algorithm that SageMaker doesn't support? For example, instead of blazing text, what if you want to use BERT models? In that scenario, we can use custom Docker images, and in SageMaker terms this is called *bring-your-own-models*. As you saw earlier, all we need is a Docker image for training and inference, training files in S3 bucket, output S3 bucket location, etc. So, to make your custom models run in SageMaker, you'll have to follow these steps:

1. Write the training script.

2. Test the training script inside the Docker container after installing all the important packages in Docker.

3. Edit the inference script inside Docker. This will be a script written probably in Python and Flask.

4. Test the inference script.

5. Once everything works fine, push the Docker image along with the training and inference script onto ECR.

6. You can now call this algorithm by passing the URL of ECR.

We will look at how to make a custom algorithm in SageMaker in the upcoming chapters.

There are a lot of other features in SageMaker that we will keep discussing throughout this book. But, for now, this introduction suffices. In the upcoming chapters we will be delving deeper into SageMaker and how to execute the code and algorithms in it.

Let's now explore some other machine learning services that AWS provides beyond SageMaker.

Other Machine Learning Services by AWS

Let's start this section by looking at Amazon Comprehend, a service dedicated to text analytics in AWS.

Amazon Comprehend

Amazon Comprehend is a service provided by AWS for NLP-related tasks. If a company has a group of documents and some extraction from them is needed so that a specific insight can be drawn out, then Amazon Comprehend is the service to use. Same tasks can be done in SageMaker as well, but Comprehend requires the least amount of coding. This is an ideal solution for people with less coding experience. The following are key elements that can be extracted from the documents:

- Entities such as organizations, places, names, etc.

- Key phrases in the documents

- The language and the sentiments of the sentences

- The syntax and syntactical structure of sentences

In addition to these features, custom classifiers can be built that can sort the document into categories based on similarity. Also, in addition to the default entities that can be extracted, custom entities can be extracted from documents. This can be done by training the base model of Amazon Comprehend. All the NLP tasks of Amazon Comprehend have a base of neural networks. That means all the tasks performed are deep learning based and can be customized as well.

Also, we can do topic modeling using Amazon Comprehend when we can find key topics present in all the documents based on the frequency and distribution of words. For this task, Amazon recommends a minimum of 1,000 documents.

Amazon Polly

Amazon Polly is the service provided by AWS for speech synthesis. Whatever text that you give to Polly, it will be converted into lifelike speech. It supports multiple languages and can be customized to the voice of our choice. The base of Amazon Polly is neural networks, just like Comprehend. It is called a *neural text-to-speech* (NTTS) model. This is the reason why Polly has the most human-like voice as compared to its competitors.

The pronouncing accuracy of Polly is super high, and it includes abbreviations, acronym expansions, and date/time interpretations; it also supports homograph disambiguation. This means that the words that have the same spellings but different meanings based on the sentence in which they are used can also be understood by Polly. This process is called *context-aware analysis*.

Polly offers both male and female voices and supports three British English voices and eight US English voices. Polly even supports voices that can sound like newscasters.

Amazon Rekognition

Amazon Rekognition is used for understanding the objects present inside an image or a video and then extracting them. The objects may include people, text, scenes, activities, inappropriate content, etc. It has a capability for facial analysis, facial comparison, and face searching. Like other services, Amazon Rekognition has a back end of deep learning with neural networks used to understand the patterns.

Amazon Rekognition's development is ongoing, which means the data is continuously updated and given to the model, with an increased number of labels. This means that the accuracy of the model keeps getting better for different categories. Some of the common use cases of Amazon Rekognition are as follows:

- Searching an image or a video for the presence of an object

- Using facial feature–based authentication

- Understanding emotional expressions such as happy, sad, enthusiastic, etc.

- Using demographic information such as gender, place, type, etc.

- Detecting adult and violent content present in videos or images

- Recognizing and extracting textual content from the images

We can also train Amazon Rekognition on custom labels, if required for a specific dataset related to a problem statement.

Amazon Translate

Machine translation is one of the applications of NLP where we translate one language to another language. Amazon has made a service totally dedicated to this use case wherein it supports neural machine translation from multiple languages to multiple languages. It is called Amazon Translate.

This service is based on an encoder-decoder architecture similar to BERT and other language models. Here, first the input language is understood using encoder architecture, and then the task of translation is done using the decoder architecture. It also uses something called an *attention layer* that tries to understand the context of a sentence by understanding the relationship between the words. Combining all these things with a neural network like long short-term memory (LSTM), the process of translation is achieved.

Amazon Transcribe

When we have multiple audio files and we want to convert whatever is spoken in a clip into text, we can use Amazon Transcribe. It has multiple applications including a combination of some other Amazon services. Here are some examples:

- A news clip can be first converted from speech to text using Transcribe, then it can be converted to your language of choice using Translate, and finally it can be read out aloud using Polly.

- Customer service calls can be recorded and transcribed. Finally, Comprehend can be used to understand different aspects of the transcribed text.

- It can be used to provide real-time subtitles.

Amazon Transcribe can be used to identify speakers based on the voice. It can be used to tell the difference between the one asking the question and the answerer, supposedly in a recorded press conference. Also, if you feel like Transcribe is not able to understand a few words, then you can increase the vocabulary of Transcribe by providing a custom vocabulary.

Amazon Textract

Amazon Textract is used to detect text present inside a document that can be in PDF format or image format. It can also extract information from tabular data, or other types of document formats that can include applications such as financial reports extraction, medical records extraction, etc. Like other services, Textract is also built on a deep neural network architecture, where the weights are updated almost daily by having access to more than a billion images and videos. Textract can also be used to extract data from forms, CSV sheets, websites, etc.

Conclusion

In this chapter, you got a machine learning–specific overview of AWS. In the next chapter, we will be looking at data processing using SageMaker and explore other services as well, which will help us in data processing.

## CHAPTER 5

Data Processing in AWS

Data processing is one of the first steps of the machine learning pipeline. As different sources of data have different formats, it becomes almost impossible to handle all the formats inside the model. Hence, we give the data a synchronous structure, and then we try to process different unwanted sections of it. These sections include the null values, outliers, dummification of categorical columns, standardization of numerical columns, etc. We can use SageMaker effectively to process the data in all these domains. This chapter assumes that you have knowledge about different data processing techniques and their implementation in Python. This chapter will be dedicated to using SageMaker to do this.

Preprocessing in Jupyter Notebook

In between receiving the raw data and feeding the data to the model, there are a lot of steps the data goes through. These steps are the data processing steps. Data processing includes feature selection, feature transformation, feature imputation, feature normalization, etc. Once all these steps are done, we proceed to splitting the data into a training set and a validation set, which are finally given to the model.

In this section, we will be looking at some of the basic data processing steps that we can follow.

1. Loading the raw data

2. Imputing the null values, which means how to replace the null values with some actual values.

3. Splitting the data into categorical and numerical data frames

4. "Dummifying" categorical data

5. Imputing the remaining null values

© Himanshu Singh 2021
H. Singh, *Practical Machine Learning with AWS*, https://doi.org/10.1007/978-1-4842-6222-1_5

6. Concatenating the categorical and numerical data frames

7. Normalizing the final data frame

8. Splitting the data into train and validation sets

This chapter assumes that you have hands-on knowledge of Pandas, Numpy, and Scikit-Learn. These packages are required for the data processing steps. If not, then it is recommended that you explore these packages to get some hands-on experience before moving on to learning SageMaker.

The dataset that we will be using for processing is the Big Mart sales dataset, which can be downloaded from Kaggle at `www.kaggle.com/devashish0507/big-mart-sales-prediction`.

This dataset contains a lot of information related to the sales of items in a retail shop. The task is to predict the sales of items. We will not be looking at the prediction part in this chapter. Rather, we'll be exploring only the data processing part of the process. Let's start by reading the train file using the Pandas framework.

```
import pandas as pd
data = pd.read_csv("Train.csv")
```

Now the entire CSV sheet's columns are saved in a data frame object named `data`. Next, let's explore the top five rows of the dataset.

```
data.head()
```

This is going to give us the output shown in Figure 5-1.

	Item_Identifier	Item_Weight	Item_Fat_Content	Item_Visibility	Item_Type	Item_MRP	Outlet_Identifier	Outlet_Establishment_Year	Outlet_Size	Outlet_Loc
0	FDA15	9.30	Low Fat	0.016047	Dairy	249.8092	OUT049	1999	Medium	
1	DRC01	5.92	Regular	0.019278	Soft Drinks	48.2692	OUT018	2009	Medium	
2	FDN15	17.50	Low Fat	0.016760	Meat	141.6180	OUT049	1999	Medium	
3	FDX07	19.20	Regular	0.000000	Fruits and Vegetables	182.0950	OUT010	1998	NaN	
4	NCD19	8.93	Low Fat	0.000000	Household	53.8614	OUT013	1987	High	

***Figure 5-1.** Top five rows of data*

As you can see, there are a lot of other columns as well that are not shown in Figure 5-1. So, let's look at the shape of the data as well as the list of all the columns.

```
print(data.shape)
print("****************************************************************")
print(data .columns)
```

This gives us the output shown in Figure 5-2.

```
(8523, 12)
****************************************************************************
Index(['Item_Identifier', 'Item_Weight', 'Item_Fat_Content', 'Item_Visibility',
       'Item_Type', 'Item_MRP', 'Outlet_Identifier',
       'Outlet_Establishment_Year', 'Outlet_Size', 'Outlet_Location_Type',
       'Outlet_Type', 'Item_Outlet_Sales'],
      dtype='object')
```

Figure 5-2. *Shape of data and columns*

As we can see, there are 8,523 rows and 12 columns. Also, we can see the names of all the columns in the list given.

As we have seen in the steps of processing, the next step is to impute the null values. So, let's take a look at all the columns that have null values.

```
data.isna().sum()
```

This code gives us the output shown in Figure 5-3.

```
Item_Identifier                0
Item_Weight                 1463
Item_Fat_Content               0
Item_Visibility                0
Item_Type                      0
Item_MRP                       0
Outlet_Identifier              0
Outlet_Establishment_Year      0
Outlet_Size                 2410
Outlet_Location_Type           0
Outlet_Type                    0
Item_Outlet_Sales              0
dtype: int64
```

Figure 5-3. *Null values exploration*

So, there are two columns with null values: Item_Weight and Outlet_Size. We can use the normal imputation methods provided by Scikit-Learn to impute these null values. But, instead, we will be using the help of nearby columns to fill in these null values. Let's look at the data types of these columns, as that is going to help us in making imputation strategies.

```
print(data['Item_Weight'].dtype)
print(data['Outlet_Size'].dtype)
```

The output shows that the Item_Weight column is a float, while the Outlet_Size column is categorical (or an object). We will first impute the Item_weight column. If we find the mean of Item_Weight and group it by Item_Type, then we can see that different item types have different means. See Figure 5-4.

```
data.groupby(['Item_Type']).mean()['Item_Weight']
```

```
Item_Type
Baking Goods              12.440901
Breads                    11.698562
Breakfast                 12.855365
Canned                    12.461479
Dairy                     13.391830
Frozen Foods              12.924729
Fruits and Vegetables     13.224769
Hard Drinks               11.664616
Health and Hygiene        13.156585
Household                 13.358193
Meat                      12.901705
Others                    13.734276
Seafood                   12.689328
Snack Foods               13.029730
Soft Drinks               12.067210
Starchy Foods             13.634060
Name: Item_Weight, dtype: float64
```

Figure 5-4. *Mean of item weight based on item type*

Looking at the output, what we can do is to impute all the null values of Item_Weight using the mean, respective of the Item_Type. This we can do by executing the following lines of code:

```
for i in data.Item_Type.value_counts().index:
    data.loc[(data['Item_Weight'].isna()) & (data['Item_Type'] == i),
    ['Item_Weight']] = \
    data.loc[data['Item_Type'] == 'Fruits and Vegetables', ['Item_
    Weight']].mean()[0]
```

Now, if we check the null values again, we get Figure 5-5.

```
Item_Identifier                  0
Item_Weight                      0
Item_Fat_Content                 0
Item_Visibility                  0
Item_Type                        0
Item_MRP                         0
Outlet_Identifier                0
Outlet_Establishment_Year        0
Outlet_Size                   2410
Outlet_Location_Type             0
Outlet_Type                      0
Item_Outlet_Sales                0
dtype: int64
```

Figure 5-5. *Removed numerical null values*

So, we successfully imputed the null values of the Item_Weight column. For Outlet_Size, what we will do next is to first split the data into numerical and categorical data frames and then impute the null values.

```
import numpy as np
cat_data = data.select_dtypes(object)
num_data = data.select_dtypes(np.number)
```

Now we have all the categorical columns in cat_data. We can check for the presence of null values again. See Figure 5-6.

```
cat_data.isna().sum()
```

```
Item_Identifier                  0
Item_Fat_Content                 0
Item_Type                        0
Outlet_Identifier                0
Outlet_Size                   2410
Outlet_Location_Type             0
Outlet_Type                      0
dtype: int64
```

Figure 5-6. *Categorical data null values*

So, the null value still exists. If we look at the categories present in the Outlet_Size columns, we will see there are three. See Figure 5-7.

```
cat_data.Outlet_Size.value_counts()
```

```
                Medium    2793
                Small     2388
                High       932
                Name: Outlet_Size, dtype: int64
```

Figure 5-7. *Categories and their counts*

But, if we look at the count of these categories based on the Outlet_Type, then it looks like Figure 5-8.

```
cat_data.groupby(['Outlet_Type','Outlet_Size']).count()
```

Outlet_Type	Outlet_Size	Item_Identifier	Item_Fat_Content	Item_Type	Outlet_Identifier	Outlet_Location_Type
Grocery Store	Small	528	528	528	528	528
Supermarket Type1	High	932	932	932	932	932
	Medium	930	930	930	930	930
	Small	1860	1860	1860	1860	1860
Supermarket Type2	Medium	928	928	928	928	928
Supermarket Type3	Medium	935	935	935	935	935

Figure 5-8. *Grouping outlet size with type*

In this figure, we can see that the maximum Small outlet size is for Grocery Store, Small for Supermarket Type1, and Medium for Supermarket Type2 and Supermarket Type3. So, we will impute the null values accordingly, based on the outlet type.

```
cat_data.loc[(cat_data['Outlet_Size'].isna()) & (cat_data['Outlet_Type'] ==
'Grocery Store'), ['Outlet_Size']] = 'Small'
cat_data.loc[(cat_data['Outlet_Size'].isna()) & (cat_data['Outlet_Type'] ==
'Supermarket Type1'), ['Outlet_Size']] = 'Small'
cat_data.loc[(cat_data['Outlet_Size'].isna()) & (cat_data['Outlet_Type'] ==
'Supermarket Type2'), ['Outlet_Size']] = 'Medium'
cat_data.loc[(cat_data['Outlet_Size'].isna()) & (cat_data['Outlet_Type'] ==
'Supermarket Type3'), ['Outlet_Size']] = 'Medium'
```

We can now check the null values for confirmation. See Figure 5-9.

```
Item_Identifier            0
Item_Fat_Content           0
Item_Type                  0
Outlet_Identifier          0
Outlet_Size                0
Outlet_Location_Type       0
Outlet_Type                0
dtype: int64
```

Figure 5-9. *All null values removed*

Finally, all the null values have been successfully removed. Remember, we can use the fillna() method of Pandas to do the same thing. Also, we can impute values using different other approaches such as backward fill, forward fill, interpolation, etc. You can experiment with all those approaches on your own.

Now that we have taken care of all the null values, we will do one last thing before moving on to dummification. If we look at the categories of the Item Fat Content column, we will see that there are the same values present in different ways. See Figure 5-10.

```
cat_data.Item_Fat_Content.value_counts()
```

```
Low Fat     5089
Regular     2889
LF           316
reg          117
low fat      112
Name: Item_Fat_Content, dtype: int64
```

Figure 5-10. *Duplicates*

LF means Low Fat, reg means Regular, and low fat is just the lowercase version of Low Fat. Let's rectify all of this.

```
cat_data.loc[cat_data['Item_Fat_Content'] == 'LF' , ['Item_Fat_Content']] =
'Low Fat'
cat_data.loc[cat_data['Item_Fat_Content'] == 'reg' , ['Item_Fat_Content']] =
'Regular'
cat_data.loc[cat_data['Item_Fat_Content'] == 'low fat' , ['Item_Fat_
Content']] = 'Low Fat'
```

Now, we will see the values shown in Figure 5-11.

```
Low Fat    5517
Regular    3006
Name: Item_Fat_Content, dtype: int64
```

Figure 5-11. *Duplicates removed*

So, this task was done successfully. Next, let's apply label encoding on the categorical data frame. We will use the Scikit-Learn package for this.

```
from sklearn.preprocessing import LabelEncoder
le = LabelEncoder()
cat_data = cat_data.apply(le.fit_transform)
```

The output results in the data frame shown in Figure 5-12.

```
cat_data.head()
```

	Item_Identifier	Item_Fat_Content	Item_Type	Outlet_Identifier	Outlet_Size	Outlet_Location_Type	Outlet_Type
0	156	0	4	9	1	0	1
1	8	1	14	3	1	2	2
2	662	0	10	9	1	0	1
3	1121	1	6	0	2	2	0
4	1297	0	9	1	0	2	1

Figure 5-12. *Label encoding output*

We will concatenate the two data frames, categorical and numerical, and then normalize the columns. Also, we will remove two of the columns before that, one in Item_Identifier and the second in Item_Sales. Item_Identifier is not really an important column, while Item_Sales will be our dependent variable; hence, it cannot be in the independent variables list. See Figure 5-13.

```
from sklearn.preprocessing import StandardScaler
ss = StandardScaler()

num_data = pd.DataFrame(ss.fit_transform(num_data.drop(['Item_Outlet_Sales'],
axis=1)), columns = num_data.drop(['Item_Outlet_Sales'],axis=1).columns)

cat_data = pd.DataFrame(ss.fit_transform(cat_data.drop(['Item_Identifier'],
axis=1)), columns = cat_data.drop(['Item_Identifier'], axis=1).columns)
```

```
final_data = pd.concat([num_data,cat_data],axis=1)
final_data.head()
```

Item_Weight	Item_Visibility	Item_MRP	Outlet_Establishment_Year	Item_Fat_Content	Item_Type	Outlet_Identifier	Outlet_Size	Outlet_Location_Type	Outlet_
-0.856325	-0.970732	1.747454	0.139541	-0.738147	-0.766479	1.507813	-0.664080	-1.369334	-0.25
-1.655730	-0.908111	-1.489023	1.334103	1.354743	1.608963	-0.607071	-0.664080	1.091569	1.00
1.083061	-0.956917	0.010040	0.139541	-0.738147	0.658786	1.507813	-0.664080	-1.369334	-0.25
1.485129	-1.281758	0.660050	0.020085	1.354743	-0.291391	-1.664513	0.799954	1.091569	-1.50
-0.943834	-1.281758	-1.399220	-1.293934	-0.738147	0.421242	-1.312032	-2.128115	1.091569	-0.25

Figure 5-13. *Standard scaling output*

Now, we have our final data ready. We have used a standard scaler class to normalize all the numerical values to their z-scores. We will be using `final_data` as independent variables, while we will extract `Item Sales` as dependent variables.

```
X = final_data
y = data['Item_Outlet_Sales']
```

The last step is to get our training and validation sets. For this we will use the class `model_selection` provided by Scikit-Learn. We will take 10 percent of our data as a validation set while remaining as a test set.

```
from sklearn.model_selection import train_test_split
X_train, X_test, y_train, y_test = train_test_split(X, y, test_size = 0.1,
random_state=5)
```

This marks the last step of data processing. Now we can use it to train any kind of model that we want. The code lines that I have shown can be executed in any Jupyter Notebook, either in the localhost or in the cloud. The only requirement is that the necessary packages must be installed.

In the next section, I will show you how to run the same code in SageMaker using the Scikit-Learn container provided by the SageMaker service. The script remains the same, but the process changes, as we have to continuously talk with the S3 bucket and define the instances as well. We will explore this in detail in the next section.

Preprocessing Using SageMaker's Scikit-Learn Container

We use SageMaker to take advantage of multiple things, especially the computation power, API generation, and ease of storage. Therefore, to achieve these things, the code must be written in a specific format. We will use the same code that we saw in the previous section, but we'll make some changes in the overall structure so that it becomes compatible with SageMaker.

First, the data should be in the S3 bucket. We have already put our Train.csv file in the bucket, in the first section of this chapter. Once that is done, we can start writing our code. First, we will define the role of the user and the region in which we are using the SageMaker service.

```
import boto3
import sagemaker
from sagemaker import get_execution_role

region = boto3.session.Session().region_name
role = get_execution_role()
```

The Boot3 package tries to extract the region name automatically if we are using the SageMaker notebook. If we are working from the localhost notebook, then it needs to be custom defined. We will look at that part in the last part of this book. get_execution_role() extracts the current role with which the user has signed in. It can be the root user or IAM role.

Now that we have defined the region and role, the next step will be to define our Scikit-Learn container. As mentioned in the first part of the book, SageMaker operates on Docker containers. All the built-in algorithms are nothing but Docker containers, and even the custom algorithm must be put inside the Docker container and uploaded to ECR. Since we will be using Scikit-Learn to process our data, already SageMaker has a processing container for that. We just need to instantiate it and then use it.

```
from sagemaker.sklearn.processing import SKLearnProcessor
sklearn_processor = SKLearnProcessor(framework_version='0.20.0',
                    role=role,
                    instance_type='ml.m5.xlarge',
                    instance_count=1)
```

In the previous code, we created an object called SKLearnProcessor. The parameters passed tell about the version of Scikit-Learn to use, the IAM role to be passed to the instance, the type of compute instance to be used, and finally the number of compute instances to be spinned up. Once this is done, any Python script that we write and that uses Scikit-Learn can be used inside this container.

Now, let's check whether our data is accessible from SageMaker.

```
import pandas as pd
input_data = 's3://slytherins-test/Train.csv'
df = pd.read_csv(input_data)
df.head()
```

slytherins-test is the name of the S3 bucket that we created earlier in the chapter. Train.csv is the data that we uploaded. If everything works perfectly, you'll get the output shown in Figure 5-14.

	Item_Identifier	Item_Weight	Item_Fat_Content	Item_Visibility	Item_Type	Item_MRP	Outlet_Identifier	Outlet_Establishment_Year	Outlet_Size	Outlet_Lo(
0	FDA15	9.300	Low Fat	0.016047	Dairy	249.8092	OUT049	1999	Medium	
1	DRC01	5.920	Regular	0.019278	Soft Drinks	48.2692	OUT018	2009	Medium	
2	FDN15	17.500	Low Fat	0.016760	Meat	141.6180	OUT049	1999	Medium	
3	FDX07	19.200	Regular	0.000000	Fruits and Vegetables	182.0950	OUT010	1998	NaN	
4	NCD19	8.930	Low Fat	0.000000	Household	53.8614	OUT013	1987	High	

Figure 5-14. *Data overview*

If you are getting any error, make sure that the bucket as well as the data has been given public access. We have talked about this in the previous part of the book.

Now, it's time to define our processing script that will be run inside the container. We have already written this script in the previous part. We will just restructure the code and save it inside a file named preprocessing.py.

```
import argparse
import os
import warnings
import pandas as pd
import numpy as np
from sklearn.model_selection import train_test_split
from sklearn.preprocessing import StandardScaler, LabelEncoder
```

```
from sklearn.exceptions import DataConversionWarning
warnings.filterwarnings(action='ignore', category=DataConversionWarning)

# Here we have defined all the columns that are present in our data
columns = ['Item_Identifier', 'Item_Weight', 'Item_Fat_Content',
'Item_Visibility','Item_Type', 'Item_MRP', 'Outlet_Identifier',
'Outlet_Establishment_Year', 'Outlet_Size', 'Outlet_Location_Type',
'Outlet_Type', 'Item_Outlet_Sales']

# This method will help us in printing the shape of our data
def print_shape(df):
    print('Data shape: {}'.format(df.shape))

if __name__=='__main__':
    # At the time of container execution we will use this parser to define
    our train validation split. Default kept is 10%
    parser = argparse.ArgumentParser()
    parser.add_argument('--train-test-split-ratio', type=float,
    default=0.1)
    args, _ = parser.parse_known_args()

    print('Received arguments {}'.format(args))

    # This is the data path inside the container where the Train.csv will
    be downloaded and saved
    input_data_path = os.path.join('/opt/ml/processing/input', 'Train.csv')

    print('Reading input data from {}'.format(input_data_path))
    data = pd.read_csv(input_data_path)
    data = pd.DataFrame(data=data, columns=columns)
    for i in data.Item_Type.value_counts().index:
        data.loc[(data['Item_Weight'].isna()) & (data['Item_Type'] == i),
        ['Item_Weight']] = \
        data.loc[data['Item_Type'] == 'Fruits and Vegetables', ['Item_
        Weight']].mean()[0]

    cat_data = data.select_dtypes(object)
    num_data = data.select_dtypes(np.number)
```

```
cat_data.loc[(cat_data['Outlet_Size'].isna()) & (cat_data['Outlet_
Type'] == 'Grocery Store'), ['Outlet_Size']] = 'Small'
cat_data.loc[(cat_data['Outlet_Size'].isna()) & (cat_data['Outlet_
Type'] == 'Supermarket Type1'), ['Outlet_Size']] = 'Small'
cat_data.loc[(cat_data['Outlet_Size'].isna()) & (cat_data['Outlet_
Type'] == 'Supermarket Type2'), ['Outlet_Size']] = 'Medium'
cat_data.loc[(cat_data['Outlet_Size'].isna()) & (cat_data['Outlet_
Type'] == 'Supermarket Type3'), ['Outlet_Size']] = 'Medium'

cat_data.loc[cat_data['Item_Fat_Content'] == 'LF' , ['Item_Fat_
Content']] = 'Low Fat'
cat_data.loc[cat_data['Item_Fat_Content'] == 'reg' , ['Item_Fat_
Content']] = 'Regular'
cat_data.loc[cat_data['Item_Fat_Content'] == 'low fat' , ['Item_Fat_
Content']] = 'Low Fat'

le = LabelEncoder()
cat_data = cat_data.apply(le.fit_transform)
ss = StandardScaler()
num_data = pd.DataFrame(ss.fit_transform(num_data), columns = num_data.
columns)
cat_data = pd.DataFrame(ss.fit_transform(cat_data), columns = cat_data.
columns)
final_data = pd.concat([num_data,cat_data],axis=1)

print('Data after cleaning: {}'.format(final_data.shape))

X = final_data.drop(['Item_Outlet_Sales'], axis=1)
y = data['Item_Outlet_Sales']

split_ratio = args.train_test_split_ratio
print('Splitting data into train and test sets with ratio {}'.
format(split_ratio))
X_train, X_test, y_train, y_test = train_test_split(X, y, test_
size=split_ratio, random_state=0)
```

101

```
# This defines the output path inside the container from where all the
csv sheets will be taken and uploaded to S3 Bucket
train_features_output_path = os.path.join('/opt/ml/processing/train',
'train_features.csv')
train_labels_output_path = os.path.join('/opt/ml/processing/train',
'train_labels.csv')
test_features_output_path = os.path.join('/opt/ml/processing/test',
'test_features.csv')
test_labels_output_path = os.path.join('/opt/ml/processing/test',
'test_labels.csv')
print('Saving training features to {}'.format(train_features_output_
path))
pd.DataFrame(X_train).to_csv(train_features_output_path, header=False,
index=False)
print('Saving test features to {}'.format(test_features_output_path))
pd.DataFrame(X_test).to_csv(test_features_output_path, header=False,
index=False)
print('Saving training labels to {}'.format(train_labels_output_path))
y_train.to_csv(train_labels_output_path, header=False, index=False)
print('Saving test labels to {}'.format(test_labels_output_path))
y_test.to_csv(test_labels_output_path, header=False, index=False)
```

As we can see, the previous code is the same as the code in the previous part of the book; all we have done is defined the place where the data will be stored inside the container and the place where the output will be stored and then uploaded to the S3 bucket from there. Once this script is defined, we are good to go now. All we have to do is spin up the instantiated container, pass this script as a parameter, pass the data as a parameter, pass the directory where output files will be stored, and finally pass the destination S3 bucket.

```
from sagemaker.processing import ProcessingInput, ProcessingOutput
sklearn_processor.run(code='preprocessing.py',
        inputs=[ProcessingInput(
        source=input_data,
        destination='/opt/ml/processing/input')],
```

```
outputs=[ProcessingOutput(output_name='train_data',
            source='/opt/ml/processing/train',
            destination='s3://slytherins-test/'),
         ProcessingOutput(output_name='test_data',
            source='/opt/ml/processing/test',
            destination='s3://slytherins-test/')],

    arguments=['--train-test-split-ratio', '0.1']
                )
```

In the previous code, we have passed all the parameters. Also, we have defined the argument that tells about the split percentage. Inside the preprocessing.py script, we have code that parses this argument.

Figure 5-15 shows what will happen next.

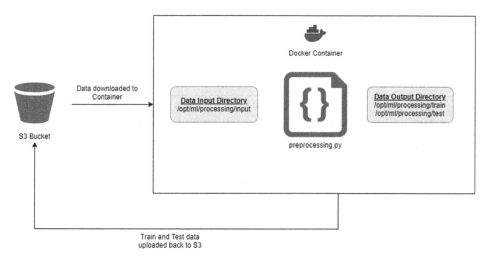

Figure 5-15. *How the processing job works*

The processing job will take some time to finish. It first launches an instance (which is similar to booting up an operating system), and then it downloads the sklearn image on the instance. Then data is downloaded to the instance. Then the processing job starts. When the job finishes, the training and test data is stored back to S3. Then the entire operation finishes. Once the job is finished, we can get detailed information about the job by using the following script:

```
preprocessing_job_description = sklearn_processor.jobs[-1].describe()
```

Let's use this script to get the S3 bucket location of the training and test datasets:

```
output_config = preprocessing_job_description['ProcessingOutputConfig']
for output in output_config['Outputs']:
    if output['OutputName'] == 'train_data':
        preprocessed_training_data = output['S3Output']['S3Uri']
    if output['OutputName'] == 'test_data':
        preprocessed_test_data = output['S3Output']['S3Uri']
```

Now, we can check the output by reading the data using Pandas.

```
training_features = pd.read_csv(preprocessed_training_data + 'train_
features.csv', nrows=10, header=None)
print('Training features shape: {}'.format(training_features.shape))
training_features.head(10)
```

This gives us the output shown in Figure 5-16.

```
Training features shape: (10, 11)
```

	0	1	2	3	4	5	6	7	8	9	10
0	0.071924	4.223950	-0.568970	-1.532846	0.047385	-0.738147	1.371418	-0.254590	0.799954	-1.369334	-1.508289
1	-0.619814	0.075491	1.969280	0.736822	0.886662	-0.738147	-0.766479	0.450371	0.799954	-0.138882	-0.252658
2	0.751946	-0.350031	-0.232154	1.095190	-1.366252	1.354743	-0.528935	-0.959551	0.799954	-0.138882	-0.252658
3	0.071924	-0.335116	-1.224896	-1.532846	-1.695729	-0.738147	1.608963	0.097891	-0.664080	1.091569	2.258603
4	0.964806	1.359713	0.480442	1.334103	-1.145858	-0.738147	-0.291391	-0.607071	-0.664080	1.091569	1.002972
5	1.603384	-0.248602	-1.329660	0.139541	-1.339537	-0.738147	-0.291391	1.507813	-0.664080	-1.369334	-0.252658
6	1.165840	1.553906	-0.752339	-1.293934	-0.164104	-0.738147	-1.479112	-1.312032	-2.128115	1.091569	-0.252658
7	1.556082	-0.977235	0.656289	-1.293934	1.447664	-0.738147	0.421242	-1.312032	-2.128115	1.091569	-0.252658
8	1.319572	-0.075335	0.077869	0.736822	-0.633832	1.354743	-0.528935	0.450371	0.799954	-0.138882	-0.252658
9	-1.623801	-0.786506	0.281015	1.095190	1.064758	-0.738147	-0.291391	-0.959551	0.799954	-0.138882	-0.252658

Figure 5-16. *Processed data*

This finishes the entire processing job that can be done using SklearnProcessor. The next step will always be to define the algorithm for machine learning. We will look at that in the next chapters.

But suppose instead of using a predefined container by SageMaker, like ScriptProcessor, we want to make our own container and run a script on that. In that case, we can use a class of SageMaker called ScriptProcessor. Let's explore that in the next section.

Creating Your Own Preprocessing Code Using ScriptProcessor

In the previous section, we used `SkLearnProcessor`, which is a built-in container provided by SageMaker. But, many times, we have to write some code that cannot only be executed in a SageMaker's predefined containers. For that we have to make our own containers. We will be looking at making our own containers while training a machine learning model as well. In this section, we will make a container that performs the same tasks as the `SKlearnProcessor` container. The only difference is that it's not prebuilt; we will build it from scratch.

To use custom containers for processing jobs, we use a class provided by SageMaker named `ScriptProcessor`. Before giving inputs to `ScriptProcessor`, the first task is to create our Docker container and push it to ECR.

Creating a Docker Container

For this we will be creating a file named `Dockerfile` with no extension. Inside this we will be downloading an image of a minimal operating system and then install our packages inside it. So, our minimal operating system will be Linux based, and we will have Python, Scikit-Learn, and Pandas installed inside it.

```
FROM python:3.7-slim-buster

RUN pip3 install pandas==0.25.3 scikit-learn==0.21.3
ENV PYTHONUNBUFFERED=TRUE

ENTRYPOINT ["python3"]
```

The previous script must be present inside the `Dockerfile`. The first line, `FROM python:3.7-slim-buster`, tells about the minimal operating system that needs to be downloaded from Docker Hub. This only contains Python 3.7 and the minimal packages required to run Python. But, we need to install other packages as well. That's why we will use the next line, `RUN pip3 install pandas==0.25.3 scikit-learn==0.21.3`. This will install Pandas, Scikit-Learn, Numpy, and other important packages. The next line, `ENV PYTHONUNBUFFERED=TRUE`, is an advanced instruction that tells Python to log messages immediately. This helps in debugging purposes. Finally, the last line, `ENTRYPOINT ["python3"]`, tells about how our `preprocessing.py` file should execute.

Building and Pushing the Image

Now that our Docker file is ready, we need to build this image and then push it to Amazon ECR, which is a Docker image repository service. To build and push this image, the following information will be required:

- Account ID

- Repository name

- Region

- Tag given to the image

All this information can be initialized using the following script:

```
import boto3

account_id = boto3.client('sts').get_caller_identity().get('Account')
ecr_repository = 'sagemaker-processing-container'
tag = ':latest'
region = boto3.session.Session().region_name
```

Once we have this information, we can start the process by first defining the ECR repository address and then executing some command-line scripts.

```
processing_repository_uri = '{}.dkr.ecr.{}.amazonaws.com/{}'.
format(account_id, region, ecr_repository + tag)

# Create ECR repository and push docker image
! docker build -t $ecr_repository docker # This builds the image
! $(aws ecr get-login --region $region --registry-ids $account_id --no-
include-email) # Logs in to AWS
! aws ecr create-repository --repository-name $ecr_repository # Creates ECR
Repository
! docker tag {ecr_repository + tag} $processing_repository_uri # Tags the
image to differentiate it from other images
! docker push $processing_repository_uri # Pushes image to ECR
```

If everything works fine, then your image will successfully be pushed to ECR. You can go to the ECR service and check the repository. You can see the view in Figure 5-17.

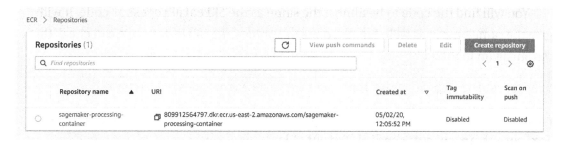

Figure 5-17. *Image pushed to ECR*

Using a ScriptProcessor Class

Now that our image is ready, we can start using the ScriptProcessor class. We will execute the same code, preprocessing.py, inside this container. Just like how we did in SKLearnProcessor, we will create an object of the class first.

```
from sagemaker.processing import ScriptProcessor, ProcessingInput,
ProcessingOutput
from sagemaker import get_execution_role

role = get_execution_role()

script_processor = ScriptProcessor(command=['python3'],
        image_uri=processing_repository_uri,
        role=role,
        instance_count=1,
        instance_type='ml.m5.xlarge')
```

Once the object is created, we can use it to run our preprocessing.py file.

```
input_data = 's3://slytherins-test/Train.csv'

script_processor.run(code='preprocessing.py',
        inputs=[ProcessingInput(
          source=input_data,
          destination='/opt/ml/processing/input')],
        outputs=[ProcessingOutput(source='/opt/ml/processing/train',
        destination='s3://slytherins-test/'),
          ProcessingOutput(source='/opt/ml/processing/test',
          destination='s3://slytherins-test/')])
```

107

You will find the code to be almost the same as the SKLearnProcessor code. It will give the same output as well. Finally, once the processing job is done, we can check the output again in the same way.

```
preprocessing_job_description = script_processor.jobs[-1].describe()

output_config = preprocessing_job_description['ProcessingOutputConfig']
for output in output_config['Outputs']:
    if output['OutputName'] == 'output-1':
        preprocessed_training_data = output['S3Output']['S3Uri']
    if output['OutputName'] == 'output-2':
        preprocessed_test_data = output['S3Output']['S3Uri']

import pandas as pd
training_features = pd.read_csv(preprocessed_training_data + 'train_
features.csv', nrows=10, header=None)
print('Training features shape: {}'.format(training_features.shape))
training_features.head(n=10)
```

The output that you'll get will be the same as shown in Figure 5-16.

In this section, we saw how we can create our own containers and run processing scripts. This becomes important in many situations. For example, if we want to use BERT-based preprocessing on an NLP task, we will have to create a container for that, as SageMaker doesn't provide us with BERT-based services. We will be exploring more about custom containers while creating training and inference jobs in later chapters.

In the past two sections, we have worked on the Jupyter Notebook inside the SageMaker container. But, most of the time, especially during production, we have to run the code in a different system. For that we will have to use the Boto3 API for the authentication and execution. In the next section, we will see how to use Boto3 for running our custom script.

Using Boto3 to Run Processing Jobs

As mentioned, we use the Boto3 package to access the services of AWS from any other computer, including your localhost. So, in this section, we will be running the custom Docker container script that we saw in the previous section, using Boto3.

Installing Boto3

The first step for using Boto3 is to install it inside the localhost environment. Along with Boto3, we have to install awscli, which will help us in authentication with AWS and s3fs, which in turn will help us in talking with the S3 bucket. To install it, we will be using pip, as shown here:

```
pip install boto3
pip install awscli
pip install s3fs
```

Once the installation finishes, we need to configure the credentials of AWS. For this, we will run the following command:

```
aws configure
```

This will ask you for the following four inputs:

- AWS access key

- AWS secret access key

- Default region name

- Default output format

Once we provide this information, we can easily use Boto3 to connect with the AWS services. I have already shown you how to get the access key and secret access key when creating the IAM roles. The default region name will be us-east-2, but you can recheck this by looking at the top-right corner of your AWS management console. It will tell you the location. As you can see in Figure 5-18, I have passed in the required information.

```
(Python_Simple) C:\Users\himan>aws configure
AWS Access Key ID [None]: AKI          D307Q
AWS Secret Access Key [None]: JbgcuUrdR                    J2sEuG
Default region name [None]: us-east-2
Default output format [None]: json
```

Figure 5-18. *Setting AWS credentials*

Once this part is done, we can start our Jupyter Notebook (local system notebook) and create a notebook using the same environment inside which we have installed all the packages and configured AWS.

Initializing Boto3

Inside the notebook, the first step will be to initialize Boto3. For this we will use the following script:

```
import boto3
import s3fs
region = boto3.session.Session().region_name
client = boto3.client('sagemaker')
```

In the previous step, the region was set up by you. The same region will be extracted and stored inside the variable region. Next is to set up Boto3. Boto3 can be set up for all the services of AWS. Currently, we will be using SageMaker; hence, we will call the class client of Boto3 and initialize it with SageMaker (`client = boto3.client('sagemaker')`).

Making Dockerfile Changes and Pushing the Image

Now, we will use the Boto3 API to call the processing job method. This will create the same processing job that we saw in the previous section. But, minor changes will be required, and we will explore them one by one.

We will use the method `create_processing_job` to run the data processing job. To learn more about this method, or all the methods related to SageMaker provided by Boto3, you can visit `https://boto3.amazonaws.com/v1/documentation/api/latest/reference/services/sagemaker.html`.

But, before that, we have to make some changes in our Docker container and our processing Python file. For the Docker container, we will need to copy our `preprocessing.py` script inside it so that the Boto3 method can run the script directly. For this we will make the following changes to our Dockerfile:

```
FROM python:3.7-slim-buster
RUN pip3 install pandas==0.25.3 scikit-learn==0.21.3
ENV PYTHONUNBUFFERED=TRUE
ENV PATH="/opt/ml/code:${PATH}"
COPY preprocessing.py /opt/ml/code/preprocessing.py
WORKDIR /opt/ml/code
```

We have added three new lines to our existing Dockerfile. The line ENV PATH="/opt/ml/code:${PATH}" sets up the environment path to /opt/ml/code. We will be placing our script, preprocessing.py, inside it with COPY preprocessing.py /opt/ml/code/preprocessing.py. Finally, we will be making our working directory the same folder: WORKDIR /opt/ml/code. This is required so that the Docker container will know where the script file is present, and it will help in its execution.

Once we have made changes in the Dockerfile, we will make changes to the script that builds the image and pushes it to the ECR. The only change that we need to do is add a line that gives the permission to the container to play with the preprocessing.py script. Otherwise, Docker may not have the permission to open and look at its contents.

```
# Create ECR repository and push docker image
! chmod +x docker/preprocessing.py # This line gives read and write access
to the preprocessing script
! docker build -t $ecr_repository docker # This builds the image
! $(aws ecr get-login --region $region --registry-ids $account_id --no-
include-email) # Logs in to AWS
! aws ecr create-repository --repository-name $ecr_repository # Creates ECR
Repository
! docker tag {ecr_repository + tag} $processing_repository_uri # Tags the
image to differentiate it from other images
! docker push $processing_repository_uri # Pushes image to ECR
```

Once this step is done, we will be ready to run our Boto3 processing job.

Creating a Processing Job

In a nutshell, we need information about four sections to create a processing job using Boto3.

- Input data information (ProcessingInput)

- Output data information (ProcessingOutput)

- Resource information (ProcessingResources)

- Container information (AppSpecification)

As you can see in the following code, all the previous information is provided. The code is again similar to the code we saw in the previous section; it is just that Boto3 needs information that should be manually put inside it as parameters, while when we run the code from inside SageMaker, most of the information is automatically extracted.

```python
response = client.create_processing_job(        # Initialize the method
    ProcessingInputs=[
        {
            'InputName': "Training_Input",      # Give Input Job a name
            'S3Input': {
                'S3Uri': input_data,            # URL from where the data
                                                  needs to be taken
                'LocalPath': '/opt/ml/processing/input',
                # Local directory where the data will be downloaded
                'S3DataType': 'S3Prefix',       # What kind of Data is it?
                'S3InputMode': 'File'           # Is it a file or a
                                                  continuous stream of data?
            }
        },
    ],
    ProcessingOutputConfig={
        'Outputs': [
            {
                'OutputName': 'Training',       # Giving Output Name
                'S3Output': {
                    'S3Uri': 's3://slytherins-test/',
                    # Where the output needs to be stored
                    'LocalPath': '/opt/ml/processing/train',
                    # Local directory where output needs to be searched
                    'S3UploadMode': 'EndOfJob'      # Upload is done when
                                                      the job finishes
                },
                'OutputName': 'Testing',
                'S3Output': {
                    'S3Uri': 's3://slytherins-test/',
                    'LocalPath': '/opt/ml/processing/test',
```

```
                'S3UploadMode': 'EndOfJob'
            }
        },
    ],
},
ProcessingJobName='preprocessing-job-test',    # Giving a name to the
                                                 entire job. It should
                                                 be unique

ProcessingResources={
    'ClusterConfig': {
        'InstanceCount': 1,                     # How many instances are
                                                  required?

        'InstanceType': 'ml.m5.xlarge',         # What's the instance
                                                  type?

        'VolumeSizeInGB': 5                     # What should be the
                                                  instance size?

    }
},
AppSp {
    'ImageUri': '809912564797.dkr.ecr.us-east-2.amazonaws.com/
    sagemaker-processing-container:latest',
# Docker Image URL
    'ContainerEntrypoint': [
        'Python3','preprocessing.py'           # How to run the script
    ]
},
RoleArn='arn:aws:iam::809912564797:role/sagemaker-full-accss',
# IAM role definition
)
```

RoleArn defines the IAM role that will be needed to run the code. We have already made this role in the activity section. I also explained how to copy the ARN during IAM role creation.

The previous code will start the processing job. But, you will not see any output. To know the status of the job, you can use CloudWatch, which I will talk about in the next section. For now, we will get help from the Boto3 method `describe_processing_job` to get the information. We can do this by writing the following code:

```
client.describe_processing_job(
    ProcessingJobName='processing-job-test'
)
```

This will give us detailed information about the job, as shown in Figure 5-19.

```
{'ProcessingInputs': [{'InputName': 'Training_Input',
   'S3Input': {'S3Uri': 's3://slytherins-test/Train.csv',
    'LocalPath': '/opt/ml/processing/input',
    'S3DataType': 'S3Prefix',
    'S3InputMode': 'File',
    'S3DataDistributionType': 'FullyReplicated'}}],
 'ProcessingOutputConfig': {'Outputs': [{'OutputName': 'Testing',
    'S3Output': {'S3Uri': 's3://slytherins-test/',
     'LocalPath': '/opt/ml/processing/test',
     'S3UploadMode': 'EndOfJob'}}]},
 'ProcessingJobName': 'preprocessing-14',
 'ProcessingResources': {'ClusterConfig': {'InstanceCount': 1,
   'InstanceType': 'ml.m5.xlarge',
   'VolumeSizeInGB': 5}},
 'StoppingCondition': {'MaxRuntimeInSeconds': 86400},
 'AppSpecification': {'ImageUri': '809912564797.dkr.ecr.us-east-2.amazonaws.com/sagemaker-processing-container:latest',
  'ContainerEntrypoint': ['python3', 'preprocessing.py']},
 'RoleArn': 'arn:aws:iam::809912564797:role/sagemaker-full-accss',
 'ProcessingJobArn': 'arn:aws:sagemaker:us-east-2:809912564797:processing-job/preprocessing-14',
 'ProcessingJobStatus': 'Completed',
 'ProcessingEndTime': datetime.datetime(2020, 5, 3, 18, 13, 43, tzinfo=tzlocal()),
 'ProcessingStartTime': datetime.datetime(2020, 5, 3, 18, 13, 26, tzinfo=tzlocal()),
 'LastModifiedTime': datetime.datetime(2020, 5, 3, 18, 13, 43, 462000, tzinfo=tzlocal()),
 'CreationTime': datetime.datetime(2020, 5, 3, 18, 10, 53, 273000, tzinfo=tzlocal()),
 'ResponseMetadata': {'RequestId': 'abc2aded-fec3-456b-986a-8714153c9a81',
  'HTTPStatusCode': 200,
  'HTTPHeaders': {'x-amzn-requestid': 'abc2aded-fec3-456b-986a-8714153c9a81',
   'content-type': 'application/x-amz-json-1.1',
   'content-length': '1104',
   'date': 'Sun, 03 May 2020 12:44:19 GMT'},
  'RetryAttempts': 0}}
```

Figure 5-19. *Processing job description*

You will find the key `ProcessingJobStatus`, which tells about the status, and if the job fails, you will get a reason for the failure key as well. So, now we have seen the three ways of data processing provided by SageMaker. Let's explore how we can monitor these jobs in the next section.

Monitoring Processing Jobs Using CloudWatch

CloudWatch is an amazing service provided by Amazon that helps you monitor almost every job, be it training, inference, or processing jobs. In this section, we will be looking at the usage of CloudWatch to monitor processing Jobs. In later chapters, we will explore it for other machine learning techniques as well.

First, once we log in to the AWS Management Console, we must go to Services and then search for *CloudWatch* and open it. Then look for the Logs section in the panel on the right and click it. See Figure 5-20.

Figure 5-20. *CloudWatch menu*

Here you can see all the log groups, depending upon the AWS services that we have used. Since we have used only two services so far, SageMaker and Processing, you'll easily find the information, as shown in Figure 5-21.

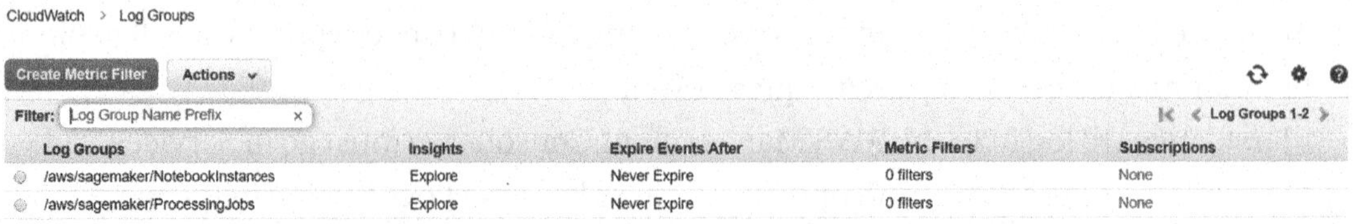

Figure 5-21. *Jobs information in CloudWatch*

We will click the ProcessingJobs section and search for the processing job name that we gave to our job. Once we find it, click the link. It will give us some output similar to Figure 5-22.

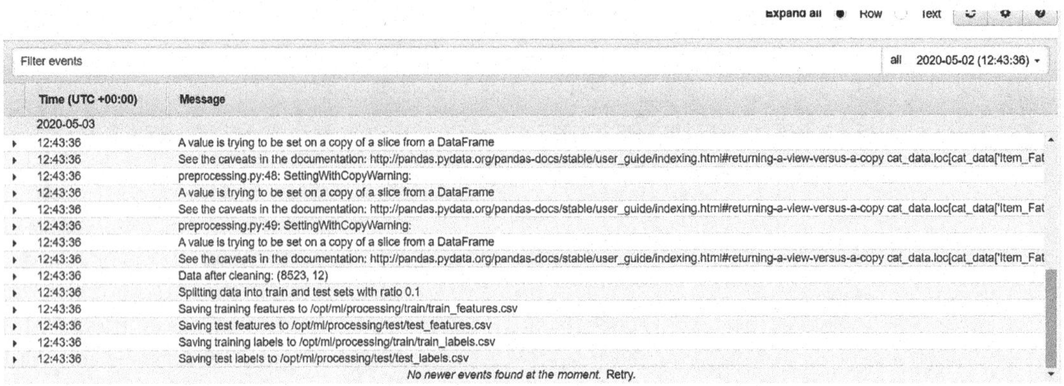

Figure 5-22. *Job logs in CloudWatch*

If you have any errors, you can find them listed here as well, based on the processing job's name. That's why I mentioned before that the name should be unique. There are a lot of other sections as well in CloudWatch, but for now they are not important. We will explore them when the need arises.

One thing to remember here is that inside the SageMaker console, you won't find the logs of the processing jobs. That's why you have to come to CloudWatch to find the job. For most training jobs, transformation jobs, etc., you'll find the logs directly in the console of SageMaker—but not for processing jobs.

Conclusion

This chapter was all about the processing of raw data using SageMaker. In the next chapter, we will look at most of the built-in algorithms of SageMaker in detail. We will start with the processing of raw data and then move on to training the model and saving the model artifacts to an S3 bucket.

CHAPTER 6

Building and Deploying Models in SageMaker

In this chapter, we will be exploring some of SageMaker's built-in algorithms that are widely used in the industry. We will be exploring the algorithms from the general domain, natural language processing domain, computer vision domain, and forecasting domain.

Exploring the Linear Learner Algorithm

The linear learner algorithm of SageMaker is similar to the regression algorithms in the machine learning domain. We can make multiple linear regression, logistic regression, and multinomial logistic regression models using the linear learner algorithm. In this section, we will look at how this algorithm can be used for linear regression and logistic regression. We will use the Big Mart dataset that we used in the previous chapter to apply this algorithm. Before delving into how to apply linear learner in SageMaker, let's take a brief look at linear and logistic regression.

Overview of Linear Regression

Linear regression is one of the most basic yet most important algorithms in machine learning. It is used to fit a line (or a curve in the case of nonlinear regression) on the observations and then interpolate the fitted line to get the predictions. To fit the line, we use an approach that is called *least squares estimations*, which gives us our coefficient values. These coefficient values are determined in such a way that the mean of the errors is approximately zero. Errors are the Euclidean distance of each observation from the fitted line. Figure 6-1 shows simple linear regression used to fit a line.

© Himanshu Singh 2021
H. Singh, *Practical Machine Learning with AWS*, https://doi.org/10.1007/978-1-4842-6222-1_6

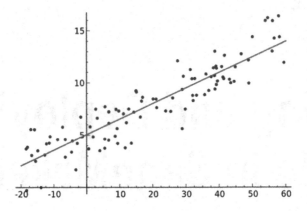

Figure 6-1. *Simple linear regression*

Linear regression is used to predict numerical values. There are various versions of regression, namely, ridge regression, lasso regression, elastic net regression, Gaussian regression, etc.

Overview of Logistic Regression

Logistic regression is the transformation of linear regression in such a way that the range of prediction is from 0 to 1. This is done by passing the equation of linear regression to a sigmoid function. Therefore, the straight line that we saw in linear regression gets converted into an S-shaped curve with an upper limit of 1 and a lower limit of 0, as shown in Figure 6-2.

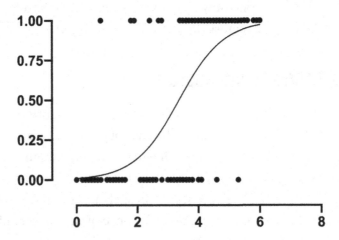

Figure 6-2. *Logistic regression*

Once logistic regression predicts a value, it is taken as a probability of prediction for binary classification use cases. As a default, if prediction exceeds 0.5, then a success class is predicted; otherwise, a failure class is predicted. The threshold of 0.5 is customizable so that we can attain better precision and recall (which will be discussed in detail in the next chapter). Logistic regression for multiclass classification is called *multinomial logistic regression*, and instead of using sigmoid, it uses the softmax function.

SageMaker Application of Linear Learner

The first step will be to read the dataset from the S3 bucket, preprocess the columns to remove the null values, and apply scaling and encoding. We saw how to preprocess the dataset and get to the dependent and independent variables in the previous chapter. Therefore, we will start this section directly by applying the algorithm on the preprocessed dataset. We will define the role and buckets so that SageMaker can talk to different services properly.

```
import boto3
from sagemaker import get_execution_role
bucket = 'slytherins-test'
prefix = 'linear-learner'
role = get_execution_role()
```

Now, we need to decide what algorithm needs to be applied, that is, linear or logistic regression. We will start with logistic regression. To make a logistic regression model, we need a categorical column. We know that our target variable is Sales, and it is a numerical column; hence, logistic regression cannot be applied. So, we will bin the Sales columns into four categories, and then we can start applying algorithms.

```
y_binned = pd.cut(y['Item_Outlet_Sales'], 4, labels=['A', 'B', 'C', 'D'])
```

The previous code bins y into the four categories of A, B, C, and D, each having an equal range. You can see the output here:

```
0        B
1        A
2        A
3        A
4        A
         ..
8518     A
8519     A
8520     A
8521     A
8522     A
Name: Item_Outlet_Sales, Length: 8523, dtype: category
Categories (4, object): [A < B < C < D]
```

Now that we have our categorical column as a target variable, we will apply label encoding on it so that each category can be represented by an integer.

```
from sklearn.preprocessing import LabelEncoder
le = LabelEncoder()
temp = le.fit(y_binned)
y_final = temp.transform(y_binned)
```

Now that we have our final target variable defined and stored in y_final, we will use it to train the model. As mentioned in the previous chapter, SageMaker runs the algorithm inside Docker containers, and hence the data should be stored in an S3 bucket so that the containers can access them. Our next step will be to store the data in S3. For our linear learner algorithm, we will use a data format called the RecordIO-Protobuf format. Using this data format helps you with a faster training time, and you can train models in live data mode (called *pipe mode*). We can convert our independent and target variables to RecordIO format using the following lines of code:

```
import io
import numpy as np
import sagemaker.amazon.common as smac

vectors = np.array(X.values, dtype='float32')
labels = np.array(y_final, dtype='float32')

buf = io.BytesIO()
smac.write_numpy_to_dense_tensor(buf, vectors, labels)
buf.seek(0)
```

The previous lines convert the data into RecordIO format and then open the temporary file so that it can be directly inserted into S3. A RecordIO file is used by breaking a big file into chunks and then using these chunks for analysis. This file helps us create streaming jobs in SageMaker, which makes the training fast. To send it, we will use the next lines of code:

```
key = 'recordio-pb-data'
boto3.resource('s3').Bucket(bucket).Object(os.path.join(prefix, 'train', key)).upload_fileobj(buf)
s3_train_data = 's3://{}/{}/train/{}'.format(bucket, prefix, key)
print('uploaded training data location: {}'.format(s3_train_data))
```

This will upload the data to S3 and close the buffer that we created. Now, our basic steps are done. All we need to do is to make the connection and train the model. The first step will be to initialize our linear learner algorithm Docker container.

```
from sagemaker.amazon.amazon_estimator import get_image_uri
container = get_image_uri(boto3.Session().region_name, 'linear-learner')
```

After initializing, let's pass the required parameters for linear learner and initialize the algorithm.

```
sess = sagemaker.Session()

linear = sagemaker.estimator.Estimator(container,
                    role,
                    train_instance_count=1,
                    train_instance_type='ml.m4.xlarge',
                     output_path=output_location,
                     sagemaker_session=sess)
```

As we know, the regression algorithms have a few hyperparameters that need to be defined, such as the number of variables, batch size, etc. We will next define these values.

```
linear.set_hyperparameters(feature_dim=11,
            predictor_type='multiclass_classifier',
            mini_batch_size=100,
            num_classes=4)
```

As everything is defined now, next we will start the training.

```
linear.fit({'train': s3_train_data})
```

You will see the output given next at the start, and then the training will start.

```
2020-05-16 17:35:45 Starting - Starting the training job...
2020-05-16 17:35:47 Starting - Launching requested ML instances......
2020-05-16 17:36:48 Starting - Preparing the instances for training...
2020-05-16 17:37:23 Downloading - Downloading input data...
2020-05-16 17:38:02 Training - Training image download completed. Training in progress.Docker entrypoint called with argu
ment(s): train
Running default environment configuration script
```

It will take some time for the model to be trained. Once the model is trained, we can deploy the model as an endpoint, and then we can start the testing. To deploy the model, we will use the deploy function.

```
linear_predictor = linear.deploy(initial_instance_count=1,
                    instance_type='ml.m4.xlarge')
```

It will take some time to deploy the model and then create the endpoint. Once done, we can start the prediction. To start the prediction, we have to first tell what kind of data the endpoint will be receiving. Then we will have to serialize the data. This format helps to efficiently transfer and store the data, regaining the original data perfectly. We can serialize our test data by using the following code:

```
from sagemaker.predictor import csv_serializer, json_deserializer
linear_predictor.content_type = 'text/csv'
linear_predictor.serializer = csv_serializer
linear_predictor.deserializer = json_deserializer
```

Now, whatever data we will be sending to the endpoint, it will be serialized and sent to the model. A prediction will come in a serialized manner, and then we will see the data in its original structure. To predict, we will be using the test data.

```
result = linear_predictor.predict(test_vectors[0])
print(result)
```

The previous line gives us the prediction for a single row. But if we want predictions for multiple rows, we can use the following code:

```
import numpy as np

predictions = []
for array in np.array_split(test_vectors, 100):
    result = linear_predictor.predict(array)
    predictions += [r['predicted_label'] for r in result['predictions']]

predictions = np.array(predictions)
```

The previous code takes 100 rows at a time and then stores the predictions for them in the variable predictions. We can now look at the model metrics using the following code:

```
from sklearn.metrics import precision_score, recall_score, f1_score

print(precision_score(labels, predictions, average='weighted'))
print(recall_score(labels, predictions, average='weighted'))
print(f1_score(labels, predictions, average='weighted'))
```

This will give us the following results:

```
0.7768463434210643
0.8108647189956588
0.7903353401644903
```

We can use Cloud Metrics as well to visualize different metrics for the model, but we will explore that in the next chapter.

Remember that once the endpoints are created, they will always run, until we stop them manually or through a script. After running all the previous code, our endpoint is still running. So, we'll stop it so that it will not incur us any cost.

```
sagemaker.Session().delete_endpoint(linear_predictor.endpoint)
```

The previous line stops the endpoint and deletes it. We have performed multinomial logistic regression in the previous example. We can use linear regression as well to predict numerical values. For that we will make the following changes in the previous code:

```
linear.set_hyperparameters(feature_dim=11,
                predictor_type='regression',
                mini_batch_size=100)
```

Also, don't forget to use the original target variable in the Big Mart dataset (the Sales column) and not the binned one. In the next section, we will apply the XGBoost algorithm on the same dataset and compare its performance with logistic regression.

Exploring the XGBoost Algorithm

XGBoost stands for *extreme gradient boosting*. In this section, we will first understand how a normal gradient descent algorithm works and how XGBoost makes it much more efficient. We will apply this algorithm on the multiclass classification of the Big Mart dataset. Let's start this section by taking a look at the two algorithms.

Gradient Boosting Algorithm

Boosting is a technique that comes inside the domain of ensemble trees in machine learning. In this algorithm, multiple decision trees are combined to give the final predictions. Other approaches in ensemble trees include bagging and random forests. Boosting differs from the other approaches by the way it combines multiple trees. When the first decision tree is made (generally its CART decision tree), then all the observations are given equal weight. The model, once trained, is applied on the same dataset. Then the second decision tree is made. In this decision tree, all the observations that were wrongly classified in the first decision tree are given more weight, while others are given less weight. This is done by increasing the weights of the observations that are difficult to classify while reducing the weights of all other observations. This process is repeated, and then finally the last decision tree gives us the final predictions. Therefore, it is said that each decision tree is boosted by the previous decision trees, which is why it's called *boosting*. There are different kinds of boosting approaches such as AdaBoost, gradient boost, light GBM, XGBoost, etc. See Figure 6-3.

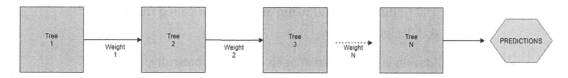

Figure 6-3. *Gradient boosting*

In gradient boosting, the weights given to the parameters are learned using the gradient descent approach. First the loss function is defined, which can be mean squared error in case of regression problems, or a logit function in the case of classification problems. This loss function is minimized after every decision tree is made and added to the next decision tree. The final aim is to minimize the overall loss function, which in return gives the best weight values for all the observations.

XGBoost Algorithm

The XGBoost algorithm is based upon the gradient boosting framework. XGBoost is a super-optimized version of gradient boosting, as it harnesses the power of computational resources so well that for small to medium size datasets, it majorly outperforms neural networks. The following are some of the major benefits of using XGBoost:

- Multithreaded operations are supported, and therefore the multiple trees that are built use a parallelization concept, and hence we can utilize large numbers of decision trees less often to give us more accurate predictions.

- Hardware performance can be maximized using approaches such as cache and buffers, out-of-core computations, etc.

- XGBoost attaches regularization factors in decision trees, and hence the problem of overfitting and underfitting is taken care of. It supports both L1 and L2 regularization.

- XGBoost automatically learns the missing values by understanding the neighborhood.

- Optimal split points are found using an approach called the *weighted quantile sketch* algorithm.

- Cross-validation is performed at each step, automatically, to find the best values of the hyperparameters. Therefore, there is no need to write custom code.

You can find a detailed overview of the XGBoost algorithm at `https://arxiv.org/pdf/1603.02754.pdf`.

SageMaker Application of XGBoost

Just like in the previous algorithm, the first step will be defining the bucket and setting up the path.

```
import os
import boto3
import re
import sagemaker
role = sagemaker.get_execution_role()
region = boto3.Session().region_name

bucket = 'slytherins-test'
prefix = 'xgboost'

bucket_path = 'https://s3-{}.amazonaws.com/{}'.format(region, bucket)
```

We will now follow the same steps of preprocessing the dataset, the steps that we saw in the previous section and the previous chapter. We will proceed from the part where we have the binned target variable. In the XGBoost algorithm, we will be using the CSV dataset, as compared to the previous one where we used RecordIO-Protobuf. We will save our data and store it in S3.

```
data_final.to_csv('train.csv', header=None, index=False)
boto3.Session(region_name=region).resource('s3').Bucket(bucket).
Object(prefix + '/train.csv').upload_file('train.csv')
```

The next step will be to initialize the Docker image of XGBoost.

```
from sagemaker.amazon.amazon_estimator import get_image_uri
container = get_image_uri(region, 'xgboost', '1.0-1')
```

Once the container is initialized, we will initialize the algorithm and run the model.

```
import boto3
from time import gmtime, strftime

job_name = 'xgboost-classification-' + strftime("%Y-%m-%d-%H-%M-%S", gmtime())

create_training_params = \
{
    "AlgorithmSpecification": {
        "TrainingImage": container,
        "TrainingInputMode": "File"
    },
    "RoleArn": role,
    "OutputDataConfig": {
        "S3OutputPath": bucket_path + "/" + prefix + "/xgboost"
    },
    "ResourceConfig": {
        "InstanceCount": 1,
        "InstanceType": "ml.m4.xlarge",
        "VolumeSizeInGB": 5
    },
    "TrainingJobName": job_name,
    "HyperParameters": {
        "max_depth":"5",
        "eta":"0.2",
        "gamma":"4",
        "min_child_weight":"6",
        "subsample":"0.7",
        "silent":"0",
        "objective":"multi:softmax",
```

```
        "num_round":"50",
        "num_class":"4"
    },
    "StoppingCondition": {
        "MaxRuntimeInSeconds": 3600
    },
    "InputDataConfig": [
        {
            "ChannelName": "train",
            "DataSource": {
                "S3DataSource": {
                    "S3DataType": "S3Prefix",
                    "S3Uri": bucket_path + "/" + prefix + '/',
                    "S3DataDistributionType": "FullyReplicated"
                }
            },
            "ContentType": "csv",
            "CompressionType": "None"
        },
        {

            "ChannelName": "validation",
            "DataSource": {
                "S3DataSource": {
                    "S3DataType": "S3Prefix",
                    "S3Uri": bucket_path + "/" + prefix + '/',
                    "S3DataDistributionType": "FullyReplicated"
                }
            },
            "ContentType": "csv",
            "CompressionType": "None"

        }
    ]
}

client = boto3.client('sagemaker', region_name=region)
client.create_training_job(**create_training_params)
```

Let's understand the previous code.

1. In the algorithm specification, we will pass the initialized Docker container and the type of data. Here we are using a CSV file; hence, the data type will be `file`.

2. In `RoleArn`, we will be passing the IAM role. It is mandatory to pass this because it will define what resources we have the access to. We can go to the IAM roles section and note the ARN of the roles that we have created there.

3. `S3OutputPath` defines where in S3 our model files will be stored.

4. Next, we have to configure our resources. We will specify the resource count, resource type, and storage. Remember, the bigger the resource you choose, the more the cost you bear. Before deciding on this, visit the Cost Explorer and look at the cost of the resource that you want to choose.

5. Hyperparameters of the XGBoost algorithm need to be set in the next section.

6. The next section talks about the maximum time you want the resource to run. If the running time exceeds that time, the job will automatically stop.

7. Finally, we pass the input data configuration and type in the last section.

This JSON format, once filled, spins up the container, and the training starts. We can get to know the metrics and logs for the model using Cloud Metrics and CloudWatch, which we will look at in the next chapter. Here we will write a script that will keep telling us whether the training is in progress. Once the training finishes or some error happens, the script informs us.

```
import time

status = client.describe_training_job(TrainingJobName=job_name)
['TrainingJobStatus']
print(status)
while status !='Completed' and status!='Failed':
    time.sleep(60)
```

```
status = client.describe_training_job(TrainingJobName=job_name)
['TrainingJobStatus']
print(status)
```

Once the training finishes, we get the output shown here:

```
Training job DEMO-xgboost-classification-2020-05-17-05-51-05
InProgress
InProgress
InProgress
InProgress
Completed
CPU times: user 92.2 ms, sys: 691 µs, total: 92.9 ms
Wall time: 4min
```

Just like in the previous section, once the model is trained, we can prepare and expose an endpoint that the predictions can use. To expose the endpoint, the following script for XGBoost algorithm can be used:

```
create_endpoint_response = client.create_endpoint(
    EndpointName="xgboost-bigmart-endpoint",
    EndpointConfigName="xgboost-bigmart-config")
```

This will spin up the endpoint, and later the predictions can be made using it. Next, we can test the model by using the invoke_endpoint() method.

```
runtime_client = boto3.client('runtime.sagemaker', region_name=region)
response = runtime_client.invoke_endpoint(EndpointName=endpoint_name,
                ContentType='text/csv',
                Body=test_data)
```

To read the predictions, we can use following script:

```
result = response['Body'].read()
result = result.decode("utf-8")
result = result.split(',')
result = [math.ceil(float(i)) for i in result]
label = payload.strip(' ').split()[0]
print ('Label: ',label,'\nPrediction: ', result[0])
```

We can do batch predictions as well, just like the linear learner. You can find the code for batch predictions in the GitHub repository. Don't forget to delete the endpoint after the predictions.

```
client.delete_endpoint(EndpointName=endpoint_name)
```

Exploring the Blazing Text Algorithm

The blazing text algorithm is used for generating word embeddings for the textual data. Later these embeddings can be given to any machine learning model to do any classification tasks. In this section, we will first understand the blazing text algorithm and then apply it on the text8 dataset.

The blazing text algorithm is a highly optimized version of the word2vec algorithm that allows faster training and inference and supports distributed training as well. Once the vectors are generated using this algorithm, we can use them for different tasks such as text classification, summarization, translation, etc. It supports two architectures, similar to that of word2vec.

- Skip gram architecture

- Continuous bag of words architecture

Let's briefly discuss these architectures.

Skip Gram Architecture of Word Vectors Generation

The skip gram algorithm is used to generate word vectors by finding words that are most similar to each other. This algorithm tries to understand the context of a sentence. To do that, it takes a word as input and then tries to predict all the words that have similar context. Figure 6-4 shows the architecture, taken from the research paper at https://arxiv.org/pdf/1301.3781.pdf (Mikolov el al.)

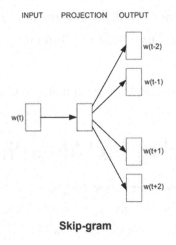

Skip-gram

Figure 6-4. *Skip gram algorithm*

To understand the context and generate word vectors, a small neural network architecture is used with hidden layers that have no activation functions. In the beginning, each word is encoded using the one-hot encoding algorithm and then fed to the network. A weight is assigned to the hidden layer, whose value is learned through a loss function. Once the model is trained, it can be used for generating word vectors or directly used for text classification models.

Continuous Bag of Words Architecture of Word Vectors Generation

The continuous bag of words (CBOW) method, you could say, is the reverse of skip gram. It understands the context and then tries to predict the word in that context. For example, if the sentence is "Delhi is the capital of India" and we then write "Delhi is the capital," then it should predict India. The architecture is again the same, where we have a hidden layer and an output layer. Each word passed to the network is one-hot encoded. See Figure 6-5.

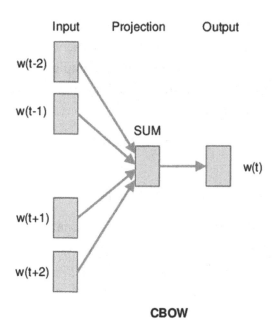

Figure 6-5. *CBOW algorithm*

SageMaker Application of Blazing Text

Before starting the coding, we must understand the dataset for which we will be generating the word vectors using the blazing text algorithm. The dataset that we'll be using is called the text8 dataset. It is a small, cleaned version of the entire Wikipedia text. The entire Wikipedia dump is called wiki9, which is then cleaned and converted into fil9. A subset (100 MB) of this cleaned dataset is taken and called text8. We can download the dataset from http://mattmahoney.net/dc/text8.zip.

As you may already know by now, the data downloaded must be sent to the S3 bucket so that our resources and the algorithm container can access it. We can upload the data using the following script:

```
train_channel = prefix + '/train'
sess.upload_data(path='text8', bucket=bucket, key_prefix=train_channel)
```

Let's store the path to this dataset in a variable.

```
s3_train_data = 's3://{}/{}'.format(bucket, train_channel)
```

Remember to define all the required fields before executing this code, namely, bucket, role, etc. We have already seen how to define them in the previous sections. We can also look at the GitHub repo to understand the complete code.

Now that we have stored the data and defined the path, the next step will be to initialize the blazing text Docker container.

```
container = sagemaker.amazon.amazon_estimator.get_image_uri(region_name,
"blazingtext", "latest")
```

Once the container is ready, we have to initialize the instance/resource.

```
bt_model = sagemaker.estimator.Estimator(container,
                        role,
                        train_instance_count=1,
                        train_instance_type='ml.m4.xlarge',
                        train_volume_size = 5,
                        train_max_run = 360000,
                        input_mode= 'File',
                        output_path=s3_output_location,
                        sagemaker_session=sess)
```

Don't forget to define the S3 output location before running this code.

```
s3_output_location = 's3://{}/{}/output'.format(bucket, prefix)
```

All the parameters are self-explanatory, and we already looked at them in the previous section. Remember, the ml.m4.xlarge instance comes under the free tier. So if you want to play around with different algorithms, always use this instance. Next, we will set up the algorithm hyperparameters.

```
bt_model.set_hyperparameters(mode="batch_skipgram",
                epochs=5,
                min_count=5,
                sampling_threshold=0.0001,
                learning_rate=0.05,
                window_size=5,
                vector_dim=100,
                negative_samples=5,
                batch_size=11,
```

```
evaluation=True,
subwords=False)
```

We have already looked at these parts of the algorithm at the start of this section. Now we will pass the data as a JSON file to train the algorithm. Before doing that, we must tell the algorithm that the data is coming from S3. This information was passed in the previous algorithm using JSON. Here we will pass it using the following script:

```
train_data = sagemaker.session.s3_input(s3_train_data, content_type='text/
plain', s3_data_type='S3Prefix')
data_channels = {'train': train_data}
bt_model.fit(inputs=data_channels, logs=True)
```

The logs parameter will not only train the model but will also show the model output in the same Jupyter Notebook. Otherwise, we would have to look at the output in the CloudWatch. The next steps will be the same as before. Deploy the model and test it.

```
bt_endpoint = bt_model.deploy(initial_instance_count = 1,instance_type =
'ml.m4.xlarge')
```

```
words = ["awesome", "blazing"]
payload = {"instances" : words}
response = bt_endpoint.predict(json.dumps(payload))
vecs = json.loads(response)
print(vecs)
```

Here we will get the output, which will be the word vectors generated for the words *awesome* and *blazing*. Finally, we will delete the model endpoint.

```
sess.delete_endpoint(bt_endpoint.endpoint)
```

In the next section, we will look at the image classification algorithm in SageMaker.

Exploring the Image Classification Algorithm

SageMaker's image classification algorithm is based upon a special convolutional neural network architecture called a ResNet. Before looking at the application of this algorithm, let's first explore and understand the ResNet architecture used for image classification.

ResNet

A ResNet is an architecture that is based on the framework of convolutional neural networks and used for problem statements such as image classification. To understand a ResNet, we must first look at the operation of convolutional neural networks. See Figure 6-6.

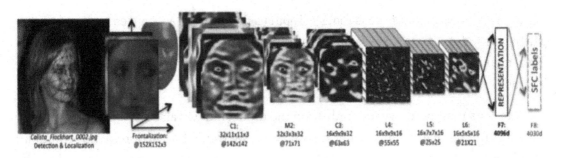

Figure 6-6. *Convolutional neural network, 10.1109/ ICEngTechnol.2017.8308186*

A typical CNN consists of the following operations:

1. The first operation is the convolution operation, which is also considered an application of filters. We apply different filters on the image so that we can get different versions of the same image, which helps us understand the image perfectly. But, instead of hard-coding the filters, the values of these filters are learned using the backpropagation approach.

2. The next step is called *pooling* or *subsampling*. Here, we reduce the size of the image so that the training time becomes faster. There are different types of pooling approaches such as max-pooling, average-pooling, etc.

3. The previous two processes are repeated multiple times, and then the final pooling operation's output is given to a fully connected neural network layer. Here the major learning happens, and finally the classification task is done.

A problem with the previous architecture is when the network is made too deep; that's when the backpropagation process suffers. Inside the backpropagation process the gradients turn to zero, and hence the learning stops. This phenomenon is called *vanishing gradients*. Therefore, to solve this issue during a deep CNN training, ResNets come into picture. Figure 6-7 shows the architecture of a ResNet.

Figure 6-7. *Source:* $https://arxiv.org/pdf/1512.03385.pdf$

ResNet's major key is that it allows the flow of gradients in the backward direction. Also, the inputs are bypassed every two convolutions. These two workarounds in CNNs solve the problem of vanishing gradients. To learn more about ResNet, please visit $https://arxiv.org/pdf/1512.03385.pdf$. Covered next is the 34-layered residual network.

SageMaker Application of Image Classification

For this algorithm, we will be using a dataset called Caltech256. It contains about 30,000 images of 256 object categories. These categories include ak47, grasshopper, bathtub, etc. We can explore more about this dataset or download the dataset from `http://www.vision.caltech.edu/Image_Datasets/Caltech256/`.

So, in this section, our task is to create a machine learning algorithm that classifies the image into these 256 categories. We will start by defining our roles, regions, etc., that we have already seen in the previous sections. Next, let's initialize the Docker container of the image classification algorithm.

```
training_image = get_image_uri(boto3.Session().region_name, 'image-classification')
```

Already we have these images categorized into train and validation sets. We can use these images directly. We can download the images from here:

```
http://data.mxnet.io/data/caltech-256/caltech-256-60-train.rec
http://data.mxnet.io/data/caltech-256/caltech-256-60-val.rec
```

Let's move these images to our S3 bucket. These images are in RecordIO-Protobuf format, as the algorithm expects them in that format only. Let's create a function this time that uploads files to S3.

```
def upload_to_s3(channel, file):
    s3 = boto3.resource('s3')
    data = open(file, "rb")
    key = channel + '/' + file
    s3.Bucket(bucket).put_object(Key=key, Body=data)
```

We will now define the folders inside the bucket where we will save the data.

```
s3_train_key = "image-classification/train"
s3_validation_key = "image-classification/validation"
```

All that is left is to store the image files in S3.

```
upload_to_s3(s3_train_key, 'caltech-256-60-train.rec')
upload_to_s3(s3_validation_key, 'caltech-256-60-val.rec')
```

Let's define the parameters related to the algorithm, which we will use to train the model.

```
num_layers = "18"
image_shape = "3,224,224"
num_training_samples = "15420"
num_classes = "257"
mini_batch_size =  "64"
epochs = "2"
learning_rate = "0.01"
```

The number of layers define the depth of the network. The image shape is 224×224 with three channels (RGB). The total number of images in the training dataset is 15,420. We have a total of 257 classes, 256 objects, and one extra class for others. We define the batch size of 64, which tells that in one go how many images will enter the network. We define the epochs as 2, which means the model will be trained on the whole training dataset two times. Finally, the learning rate is chosen as 0.1, which will decide the number of steps taken to converge and reach the local minima.

We can now define the algorithm. We have already initialized the container.

```
s3 = boto3.client('s3')

job_name_prefix = 'imageclassification'

job_name = job_name_prefix + '-' + time.strftime('-%Y-%m-%d-%H-%M-%S',
time.gmtime())

training_params = \
{
    "AlgorithmSpecification": {
        "TrainingImage": training_image,
        "TrainingInputMode": "File"
    },
    "RoleArn": role,
    "OutputDataConfig": {
        "S3OutputPath": 's3://{}/{}/output'.format(bucket, job_name_prefix)
    },
    "ResourceConfig": {
        "InstanceCount": 1,
```

```
        "InstanceType": "ml.p2.xlarge",
        "VolumeSizeInGB": 50
    },
    "TrainingJobName": job_name,
    "HyperParameters": {
        "image_shape": image_shape,
        "num_layers": str(num_layers),
        "num_training_samples": str(num_training_samples),
        "num_classes": str(num_classes),
        "mini_batch_size": str(mini_batch_size),
        "epochs": str(epochs),
        "learning_rate": str(learning_rate)
    },
    "StoppingCondition": {
        "MaxRuntimeInSeconds": 360000
    },
    "InputDataConfig": [
        {
            "ChannelName": "train",
            "DataSource": {
                "S3DataSource": {
                    "S3DataType": "S3Prefix",
                    "S3Uri": s3_train,
                    "S3DataDistributionType": "FullyReplicated"
                }
            },
            "ContentType": "application/x-recordio",
            "CompressionType": "None"
        },
        {
            "ChannelName": "validation",
            "DataSource": {
                "S3DataSource": {
                    "S3DataType": "S3Prefix",
                    "S3Uri": s3_validation,
```

```
                    "S3DataDistributionType": "FullyReplicated"
                }
            },
            "ContentType": "application/x-recordio",
            "CompressionType": "None"
        }
    ]
}
```

We already know most of the parameters in the previous JSON, as we have covered them in the XGBoost algorithm. The following are some of the unique parameters in this algorithm:

- ContentType is application/x-recordio. As I already mentioned, image classification expects only the RecordIO-Protobuf data format.

- S3DataDistributionType is fully replicated, which means if we use multiple instances for parallel training, then the dataset will be replicated in all the instances.

- The instance type we are using is p2.xlarge as image classification expects an instance having a graphics card. Be aware that the p2 and p3 instances are not at all free, and they are chargeable.

Once we are done with algorithm specifications, we will start the training process.

```
sagemaker = boto3.client(service_name='sagemaker')
sagemaker.create_training_job(**training_params)
status = sagemaker.describe_training_job(TrainingJobName=job_name)
['TrainingJobStatus']
print(status)
while status !='Completed' and status!='Failed':
    time.sleep(60)
    status = client.describe_training_job(TrainingJobName=job_name)
    ['TrainingJobStatus']
    print(status)
```

As shown, this code will start the training and then inform us whether the training successfully finished. Once the training is finished, we will deploy the model and then do the predictions. Again, the process will be the same as the ones we saw in the previous algorithms.

```
endpoint_response= sagemaker.create_endpoint(
    EndpointName="image-classification-caltech-endpoint",
    EndpointConfigName="image-classification-caltech-config")
```

This will take some time and spin up the resource required for the endpoint generation. Once the endpoint is generated, we can start the predictions.

Let's download an image and use it for testing our model. We can download a bathtub image and check whether the model predicts it perfectly.

```
! wget -O /tmp/test.jpg http://www.vision.caltech.edu/Image_Datasets/
Caltech256/images/008.bathtub/008_0007.jpg
```

The previous line will directly download the image to your system. If your system is not Linux, then you can directly go to the link and download the image. Next, we need to read the image and then pass it to the endpoint.

```
with open('/tmp/test.jpg', 'rb') as f:
    payload = f.read()
    payload = bytearray(payload)

response = runtime.invoke_endpoint(EndpointName=endpoint_name,
                    ContentType='application/x-image',
                    Body=payload)

result = response['Body'].read()
result = json.loads(result)
```

The variable result consists of probabilities of prediction for all the classes. We need to find the class that has the maximum probability. That means if we can get the argument that has the maximum probability, that will be our predicted class. For this we can use the np.argmax() function.

```
index = np.argmax(result)
```

Now, we can use this index to extract the label. We can create a list of all the labels in the same sequence as they are present in the dataset, and then we can pass the index to predict the label. Suppose we save all the classes in the variable object_classes. Next we can print the prediction.

```
print("Result: label - " + object_categories[index] + ", probability - " +
str(result[index]))
```

You can find the entire code in the GitHub repository. Also, don't forget to delete the endpoint once all the operations are done.

```
sage.delete_endpoint(EndpointName=endpoint_name)
```

Exploring the SeqToSeq Algorithm

Amazon's sequence-to-sequence algorithm is based upon recurrent neural networks, convolutional neural networks, and an encoder-decoder architecture to understand the context more efficiently. The next section is a brief overview of the RNN and encoder-decoder architectures.

Recurrent Neural Networks

When we deal with sequential data or time-based data, it becomes necessary to remember a few things from the past and understand how it can be used to predict the outcome. This is not possible with using normal artificial neural networks or convolutional neural networks. Therefore, a new architecture called RNN is used whenever we deal with sequential data. Figure 6-8 shows a simple RNN architecture.

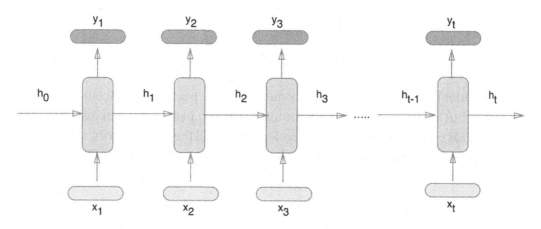

Figure 6-8. *Recurrent neural networks*

For example, in text classification, each word of the text is taken, some neural network–based computations are applied, and important aspects are stored and then passed to the next RNN layer. Storage happens in h, words are sent through x, while the output is received through y. The words are not directly passed, but they are converted into vectors and then passed. We can use algorithms such as word2vec, glove, or blazing text in SageMaker to generate these word vectors.

There are various modifications to RNNs that solve the shortcomings present in the original versions. Two of the most used are long short-term memory (LSTM) and gated recurrent units (GRU).

Encoder-Decoder Architecture

Figure 6-9 shows a typical encoder and decoder architecture.

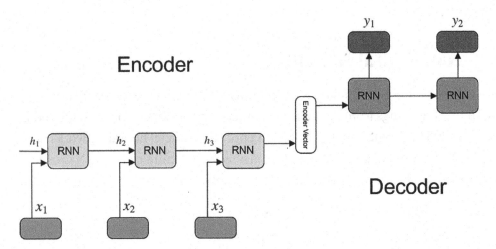

Figure 6-9. *Encoders and decoders*

An encoder is mostly used to not only memorize the past and give accurate predictions but also to understand the context of the text passed. We can use normal RNNs or LSTMS and GRUs. Once the encoders look at all the word vectors, they generate the encoder vectors and pass them to the decoder. The encoder vector suffices all the information that the encoder has received, and the decoder uses it to make efficient predictions.

The decoder takes these encoder vectors, feeds them to RNNs of its own, and then applies a softmax activation function to give the output. The best advantage of this architecture, apart from understanding the context, is its ability to take variable-length input and give variable-length output.

SageMaker Application of SeqToSeq

Let's understand the algorithm in more detail by applying it to the machine translation use case; that is, let's translate something from English to German. The first few steps will remain the same, as we saw in the previous algorithms.

```python
from time import gmtime, strftime
import time
import numpy as np
import os
import json
import boto3
import re
from sagemaker import get_execution_role
region_name = boto3.Session().region_name

bucket = 'slytherins-test'
prefix = 'seq2seq-E2G'
role = get_execution_role()

from sagemaker.amazon.amazon_estimator import get_image_uri
container = get_image_uri(region_name, 'seq2seq')
```

So, in the previous steps we have defined the container of the algorithm and defined our bucket and the folder inside where the entire model-related files will be saved. The next step will be to have a dataset. The Seq2Seq algorithm has two approaches. In the first, you can use the pretrained model available for the predictions. So, for our example, a model already exists that is trained on English to German machine translation. Or, we can train the model on our own corpus and then use it for the predictions. This process may take a lot of time, but it is the best when used for domain-specific translation tasks.

We will first see how to train the model on a corpus, and then we will use the pretrained model for predictions. The data that we will be using is news data. We will have files that contain news commentary in English and its translation in German. We can get these files from http://data.statmt.org/wmt17/translation-task/.

Let's download the data from inside the notebook and create our training and validation sets.

```
! wget http://data.statmt.org/wmt17/translation-task/preprocessed/de-en/
corpus.tc.de.gz
```

```
! wget http://data.statmt.org/wmt17/translation-task/preprocessed/de-en/
corpus.tc.en.gz
```

```
! gunzip corpus.tc.de.gz
```

```
! gunzip corpus.tc.en.gz
```

```
! mkdir validation
```

```
! curl http://data.statmt.org/wmt17/translation-task/preprocessed/de-en/
dev.tgz | tar xvzf - -C validation
```

The previous files that we have downloaded are big, around 250 MB each. So, if we train the model on the entire dataset, it may take days to finish. Therefore, we can take a subset of the entire data and use it for training.

```
! head -n 10000 corpus.tc.en > corpus.tc.en.small
! head -n 10000 corpus.tc.de > corpus.tc.de.small
```

The previous subset created has 10,000 rows. We will use this small dataset for training. The next step will be to generate English and German vocabulary from the previous files. This will use the tokenization and other NLP components to generate the vocabulary.

```
%%bash
python3 create_vocab_proto.py \
        --train-source corpus.tc.en.small \
        --train-target corpus.tc.de.small \
        --val-source validation/newstest2014.tc.en \
        --val-target validation/newstest2014.tc.de
```

The previous Python script takes as input the source English text and target German text. It applies the preprocessing to generate the vocabulary. Finally, it saves the English and German vocabulary in the validation folder. We use %%bash to run any command-line scripts inside the notebook. This is a Jupyter magic function.

Now that our dataset has been created, we need to send it to our S3 bucket.

```
def upload_to_s3(bucket, prefix, channel, file):
    s3 = boto3.resource('s3')
    data = open(file, "rb")
    key = prefix + "/" + channel + '/' + file
    s3.Bucket(bucket).put_object(Key=key, Body=data)

upload_to_s3(bucket, prefix, 'train', 'train.rec')
upload_to_s3(bucket, prefix, 'validation', 'val.rec')
upload_to_s3(bucket, prefix, 'vocab', 'vocab.src.json')
upload_to_s3(bucket, prefix, 'vocab', 'vocab.trg.json')
```

The code that we just executed generates two files. One is the vocabulary that is generated, and the second is the RecordIO-Protobuf version of the data. We will upload both of these files to S3 using the previous code.

All the basic steps are complete now, and we want to now initialize the algorithm. We will do that using the code shown here:

```
job_name = 'seq2seq-E2G'
print("Training job", job_name)

create_training_params = \
{
    "AlgorithmSpecification": {
        "TrainingImage": container,
        "TrainingInputMode": "File"
    },
    "RoleArn": role,
    "OutputDataConfig": {
        "S3OutputPath": "s3://{}/{}/".format(bucket, prefix)
    },
    "ResourceConfig": {
        # Seq2Seq does not support multiple machines. Currently, it only
        supports single machine, multiple GPUs
        "InstanceCount": 1,
        "InstanceType": "ml.m4.xlarge", # We suggest one of ["ml.
        p2.16xlarge", "ml.p2.8xlarge", "ml.p2.xlarge"]
```

```
            "VolumeSizeInGB": 5
    },
    "TrainingJobName": job_name,
    "HyperParameters": {
        # Please refer to the documentation for complete list of parameters
        "max_seq_len_source": "60",
        "max_seq_len_target": "60",
        "optimized_metric": "bleu",
        "batch_size": "64", # Please use a larger batch size (256 or 512)
        if using ml.p2.8xlarge or ml.p2.16xlarge
        "checkpoint_frequency_num_batches": "1000",
        "rnn_num_hidden": "512",
        "num_layers_encoder": "1",
        "num_layers_decoder": "1",
        "num_embed_source": "512",
        "num_embed_target": "512"
    },
    "StoppingCondition": {
        "MaxRuntimeInSeconds": 48 * 3600
    },
    "InputDataConfig": [
        {
            "ChannelName": "train",
            "DataSource": {
                "S3DataSource": {
                    "S3DataType": "S3Prefix",
                    "S3Uri": "s3://{}/{}/train/".format(bucket, prefix),
                    "S3DataDistributionType": "FullyReplicated"
                }
            },
        },
        {
            "ChannelName": "vocab",
            "DataSource": {
                "S3DataSource": {
```

```
                    "S3DataType": "S3Prefix",
                    "S3Uri": "s3://{}/{}/vocab/".format(bucket, prefix),
                    "S3DataDistributionType": "FullyReplicated"
                }
            },
        },
        {
            "ChannelName": "validation",
            "DataSource": {
                "S3DataSource": {
                    "S3DataType": "S3Prefix",
                    "S3Uri": "s3://{}/{}/validation/".format(bucket, prefix),
                    "S3DataDistributionType": "FullyReplicated"
                }
            },
        }
    ]
}

sagemaker_client = boto3.Session().client(service_name='sagemaker')
sagemaker_client.create_training_job(**create_training_params)
```

You can see that the format is the same as that of the XGBoost algorithm and image classification algorithm. All the parameters are the same as the algorithms that we discussed in the previous sections. Only the hyperparameters are specific to this algorithm. Let's discuss these parameters:

- The max sequence length of the original text and the target text is the number of characters to take as a sequence and pass to the neural network architecture.

- The batch size is the number of rows to be passed to the algorithm in one go.

- The checkpoint frequency saves the model after every batch of 1,000 rows.

- The number of hidden layers of a neural network is defined as 512 with one encoder architecture unit and one decoder architecture unit. Remember, these units are used for understanding the context of the sentences.

- The embedding source and target defines the word vector size of each sentence in the dataset. It is set to 512.

This code will start the execution of the training and will take a lot of hours to finish. Remember, this algorithm requires a GPU instance for execution. So, whatever instance you select will be chargeable. Choose wisely.

Now, let's look at how we can use the pretrained model that already exists and do the inference on the test dataset by exposing the endpoint. When we train the previous model, we will get three files:

- `Model.tar.gz`

- `Vocab.src.json`

- `Vocab.trg.json`

So, once you train the model, you can use these files directly. But, for using the pretrained model, we will download these files. We can download them from here:

```
model_name = "DEMO-pretrained-en-de-model"
! curl https://s3-us-west-2.amazonaws.com/seq2seq-data/model.tar.gz >
model.tar.gz
! curl https://s3-us-west-2.amazonaws.com/seq2seq-data/vocab.src.json >
vocab.src.json
! curl https://s3-us-west-2.amazonaws.com/seq2seq-data/vocab.trg.json >
vocab.trg.json
```

We will have to upload the model files to S3 so that our endpoint can use it.

```
upload_to_s3(bucket, prefix, 'pretrained_model', 'model.tar.gz')
model_data = "s3://{}/{}/pretrained_model/model.tar.gz".format(bucket, prefix)
```

`model_data` stores the address of the model file uploaded. Next, we will have to update this model in the algorithm so that we can use it for prediction. For this we will use the `create_model()` function.

```
sage = boto3.client('sagemaker')
primary_container = {
    'Image': container,
    'ModelDataUrl': model_data
}

create_model_response = sage.create_model(
    ModelName = model_name,
    ExecutionRoleArn = role,
    PrimaryContainer = primary_container)
```

The next step will be to define the resources that will be used by the endpoint.

```
from time import gmtime, strftime

endpoint_config_name = 'DEMO-Seq2SeqEndpointConfig-' + strftime("%Y-%m-%d-
%H-%M-%S", gmtime())
print(endpoint_config_name)
create_endpoint_config_response = sage.create_endpoint_config(
    EndpointConfigName = endpoint_config_name,
    ProductionVariants=[{
        'InstanceType':'ml.m4.xlarge',
        'InitialInstanceCount':1,
        'ModelName':model_name,
        'VariantName':'AllTraffic'}])
```

Now we can expose the endpoint by using the previous configurations.

```
endpoint_name = 'DEMO-Seq2SeqEndpoint-' + strftime("%Y-%m-%d-%H-%M-%S", gmtime())

create_endpoint_response = sage.create_endpoint(
    EndpointName=endpoint_name,
    EndpointConfigName=endpoint_config_name)
```

After some time, our endpoint will be ready for inference. Let's see how we can make predictions, in this case converting text in English to German.

```
runtime = boto3.client(service_name='runtime.sagemaker')
sentences = ["you are so good !",
```

153

```
                    "can you drive a car ?",
                    "i want to watch a movie ."
                    ]
payload = {"instances" : []}
for sent in sentences:
    payload["instances"].append({"data" : sent})

response = runtime.invoke_endpoint(EndpointName=endpoint_name,
                                   ContentType='application/json',
                                   Body=json.dumps(payload))

response = response["Body"].read().decode("utf-8")
response = json.loads(response)
print(response)
```

You will get the output as given next:

```
{'predictions': [{'target': 'Sie sind so gut !'}, {'target': 'Können Sie ein Auto fahren ?'}, {'target': 'i want to watch a
movie .'}]}
```

As you can see, the predictions have been successfully made.

Conclusion

In this chapter, you learned about the various built-in algorithms of SageMaker. These are the optimized versions of the algorithms already present in the machine learning domain. In the next chapter, we will explore different metrics with which we can evaluate these models using Cloud Metrics, look at the logs when the container is running using CloudWatch, and explore endpoint configurations in detail with connectivity with lambda functions. Also, we will do batch transformations on the algorithms that we have already seen in this chapter.

CHAPTER 7

Using CloudWatch with SageMaker

In this chapter, we will explore CloudWatch functionality in AWS in detail. Specifically, we will look at two components of CloudWatch, CloudWatch Logs and CloudWatch Metrics, that we will use a lot while using SageMaker.

Amazon CloudWatch

Amazon CloudWatch is a service provided by Amazon that tracks the resource activities of AWS and provides metrics related to it. It also stores the logs that are provided by every resource used. Through these logs and metrics, a user can explore the performance of an AWS resource being used and what can be done to improve it.

When it comes to machine learning, especially with SageMaker, CloudWatch Logs gives us the output of containers in which the code is running. As we have already seen in the previous chapters, machine learning algorithms run inside a Docker container attached to an EC2 instance. So, the output that originates from these containers is not directly visible. To look at this output, we must make some adjustments to our code, and then the status can be seen directly in the Jupyter Notebook in use, or we can use CloudWatch Logs to get this output in a step-by-step manner. The output can include your model outputs, the reason why your model failed, insights into the step-by-step execution, etc. Containers are required for three jobs, and hence we have three log groups in machine learning.

- Processing Jobs log group

- Training Jobs log group

- Transform Jobs log group

© Himanshu Singh 2021
H. Singh, *Practical Machine Learning with AWS*, https://doi.org/10.1007/978-1-4842-6222-1_7

We will look at these log groups in detail in the coming sections.

CloudWatch Metrics provides us with information in the form of values to variables. For example, when it comes to machine learning, CloudWatch Metrics can provide values such as the accuracy of a model, precision, error, etc. It can also provide metrics related to resources, such as GPU utilization, memory utilization etc. We will look at CloudWatch Metrics in detail in the coming sections. Figure 7-1 shows the architecture of how CloudWatch works.

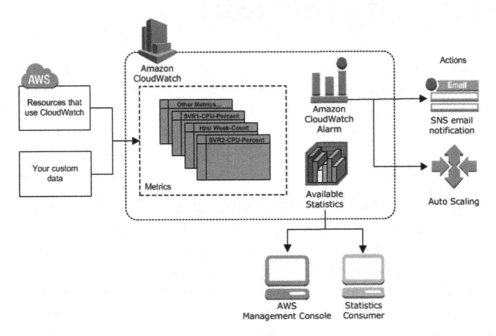

Figure 7-1. *Amazon CloudWatch architecture*

By looking at this architecture, we can see that in addition to accessing the services through the Amazon Management Console, we can integrate alarms through Amazon SNS, which can be connected to your email. We can also set custom rules based on certain criteria. This includes starting, stopping, and terminating a process or using features such as autoscaling. Let's dive deeper into CloudWatch Logs in the next section.

CloudWatch Logs

In the SageMaker console, on the left side, we have a sidebar that guides us through the different operations that are possible in it. We can create notebook instances, look at different algorithms that we ran, and analyze the endpoints. We can look at the logs of all the services that we have used by viewing the log details. Let's start with the training job–related logs.

Training Jobs

In the previous chapter, we ran an XGBoost model on the Big Mart dataset. Inside the SageMaker console, if we go to the Training drop-down and select "Training jobs" (Figure 7-2), we will get a list of all the algorithms that we have run (Figure 7-3).

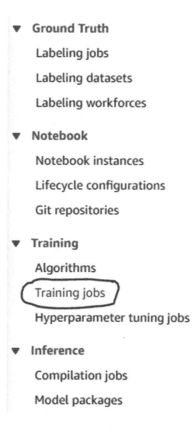

Figure 7-2. *Selecting the training job*

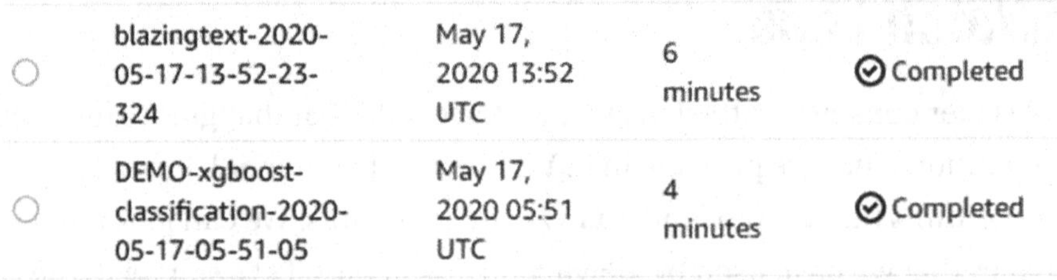

	blazingtext-2020-05-17-13-52-23-324	May 17, 2020 13:52 UTC	6 minutes	⊘ Completed
	DEMO-xgboost-classification-2020-05-17-05-51-05	May 17, 2020 05:51 UTC	4 minutes	⊘ Completed

Figure 7-3. *Selecting the job for which the logs need to be analyzed*

We can select the algorithm that we want more details for; in our case, it is XGBoost. When we click the algorithm, a new page opens with a lot of information about the model we ran. For example, it tells about the algorithm specifications that we provided while running the model. If we scroll down the page, we will come to a section called Monitor (Figure 7-4). From there we can jump to CloudWatch Logs.

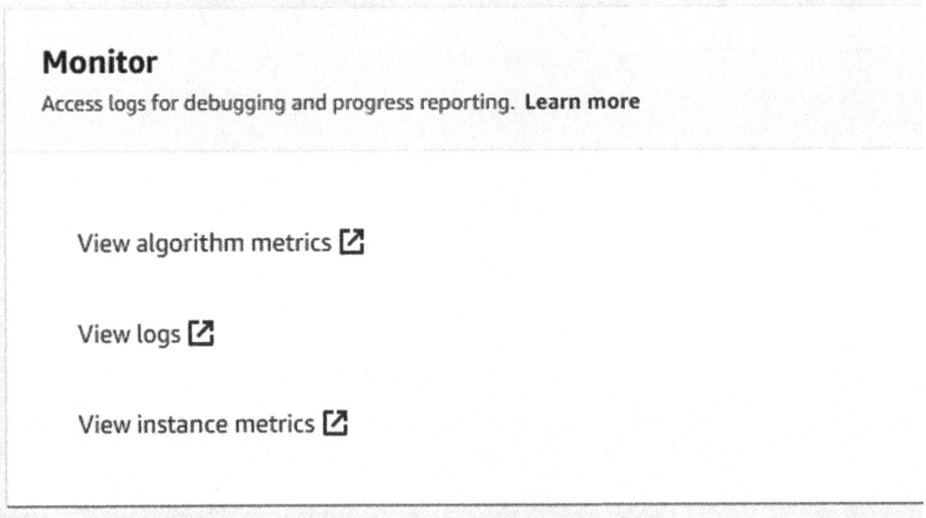

Monitor

Access logs for debugging and progress reporting. **Learn more**

View algorithm metrics ↗

View logs ↗

View instance metrics ↗

Figure 7-4. *Monitor section*

Click the "View logs" link, and you will see the CloudWatch page open. See Figure 7-5.

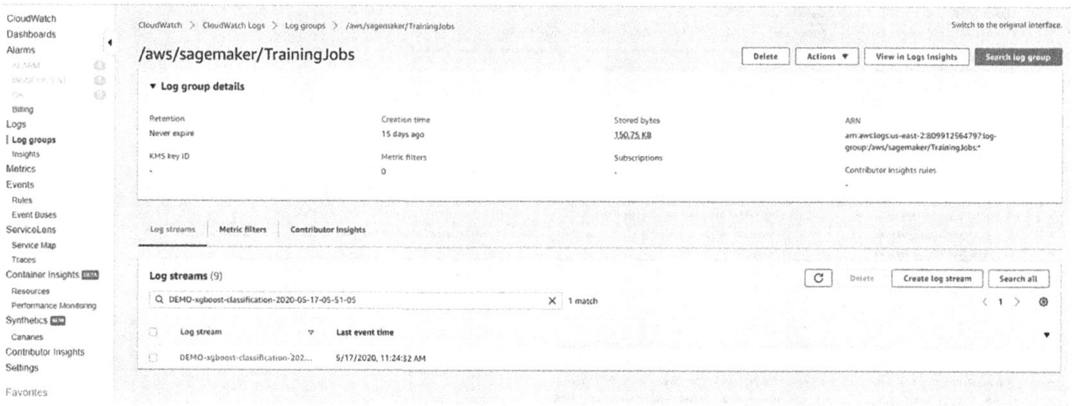

Figure 7-5. *Selecting the algorithm for the logs*

As you can see in Figure 7-5, a lot of options are available in the CloudWatch console. We are looking at the log groups. As you know, because we have selected the training job, this information is present at the top of the console (see Figure 7-6).

/aws/sagemaker/TrainingJobs

Figure 7-6. *Path*

When we scroll down, at the bottom we can see the name of the algorithm. We must click it to get to the logs (see Figure 7-7).

▶	Timestamp	Message
		There are older events to load. *Load more.*
▶	2020-05-17T11:24:32.054+05:30	INFO:sagemaker-containers:Imported framework sagemaker_xgboost_container.training
▶	2020-05-17T11:24:32.055+05:30	INFO:sagemaker-containers:Failed to parse hyperparameter objective value multi:softmax to Json.
▶	2020-05-17T11:24:32.055+05:30	Returning the value itself
▶	2020-05-17T11:24:32.055+05:30	INFO:sagemaker-containers:No GPUs detected (normal if no gpus installed)
▶	2020-05-17T11:24:32.055+05:30	INFO:sagemaker_xgboost_container.training:Running XGBoost SageMaker in algorithm mode
▶	2020-05-17T11:24:32.055+05:30	INFO:root:Determined delimiter of CSV input is ','
▶	2020-05-17T11:24:32.055+05:30	INFO:root:Determined delimiter of CSV input is ','
▶	2020-05-17T11:24:32.055+05:30	INFO:root:Determined delimiter of CSV input is ','
▶	2020-05-17T11:24:32.055+05:30	INFO:root:Determined delimiter of CSV input is ','
▶	2020-05-17T11:24:32.055+05:30	INFO:root:Determined delimiter of CSV input is ','
▶	2020-05-17T11:24:32.055+05:30	[05:54:27] 25569x11 matrix with 281259 entries loaded from /opt/ml/input/data/train?format=csv&label_column=0&delimiter=.
▶	2020-05-17T11:24:32.055+05:30	INFO:root:Determined delimiter of CSV input is ','
▶	2020-05-17T11:24:32.055+05:30	[05:54:27] 25569x11 matrix with 281259 entries loaded from /opt/ml/input/data/validation?format=csv&label_column=0&delimiter=.
▶	2020-05-17T11:24:32.055+05:30	INFO:root:Single node training.
▶	2020-05-17T11:24:32.055+05:30	INFO:root:Train matrix has 25569 rows
▶	2020-05-17T11:24:32.055+05:30	INFO:root:Validation matrix has 25569 rows
▶	2020-05-17T11:24:32.055+05:30	[05:54:27] WARNING: /workspace/src/learner.cc:328:
▶	2020-05-17T11:24:32.055+05:30	Parameters: { num_round, silent } might not be used. This may not be accurate due to some parameters are only used in language bindings but passed down to XGBoost core.
▶	2020-05-17T11:24:32.055+05:30	[0]#011train-merror:0.17916#011validation-merror:0.17916
▶	2020-05-17T11:24:32.055+05:30	[1]#011train-merror:0.17822#011validation-merror:0.17822
▶	2020-05-17T11:24:32.055+05:30	[2]#011train-merror:0.17728#011validation-merror:0.17728
▶	2020-05-17T11:24:32.055+05:30	[3]#011train-merror:0.17728#011validation-merror:0.17728
▶	2020-05-17T11:24:32.055+05:30	[4]#011train-merror:0.17658#011validation-merror:0.17658

Figure 7-7. *Visualizing the logs*

You can see that the complete logs related to your algorithm will be present there. We get the step-by-step results, which in this case includes the train and validation error. You can keep scrolling down until the end of the page to reach the last output of the algorithm. This is how you can look at the logs of any training algorithm that you have executed.

Remember, the logs will start appearing only once the container has successfully started and the algorithms have started running. If there is a problem with your Docker script, then you will not find any logs generated. But, if your code related to the algorithm has an error, then you will find the information in the logs section, as the container had successfully started, and hence the logs have started generating. So, if your model is not running, you can come to the logs to check the error in the code. With custom containers, if the logs don't start running, in general the error is probably in the Docker script. We will explore the custom containers in the next chapter. Let's now look at the logs for processing jobs.

Processing Jobs

In Chapter 5, we saw how to process data using the processing script. We used both Sklearn containers, and I also showed you how to use a custom container for processing. Let's look at the logs generated by our processing script. To do this, we will first open the CloudWatch console from the services list. Once we are there, click the Log Groups section, the one that we used in the previous section. Here, you will find a list of different log groups, as you can see in Figure 7-8.

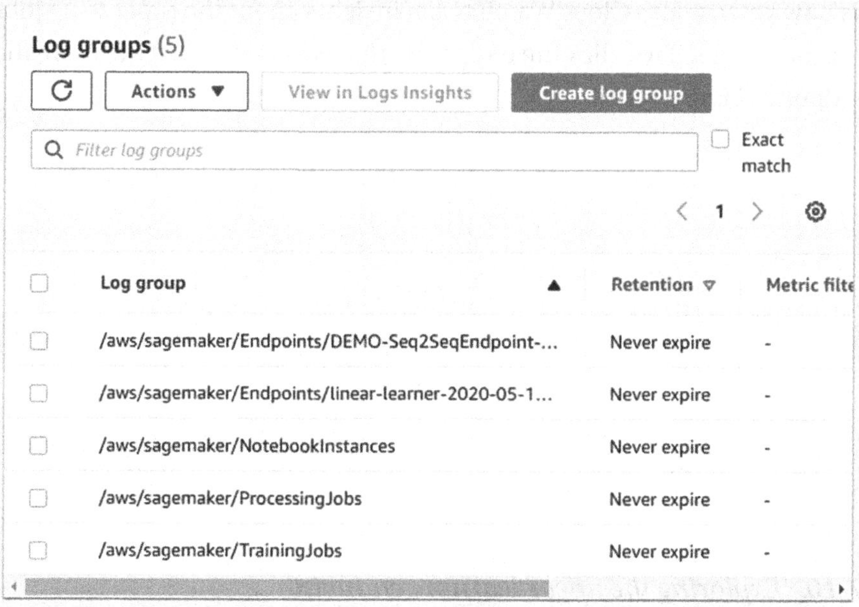

Figure 7-8. *Log groups*

You can see that in addition to the training job, you will find processing jobs, notebook instances, and endpoint logs. Let's click the processing jobs. You will find a list of all the processing jobs that you ran. If you remember the name of the job, then it will be easier to find it. That's why it is always recommended to use unique identifiers for all kinds of jobs. Click the latest processing job that we ran. Once you click it, you'll find a list of all the steps output that the processing container gave, as shown in Figure 7-9.

Figure 7-9. *Output of a container*

This is how you can use CloudWatch to get insights about the processing jobs. Let's see what output we get if we click the endpoints that we created. Let's explore the linear learner endpoint (Figure 7-10).

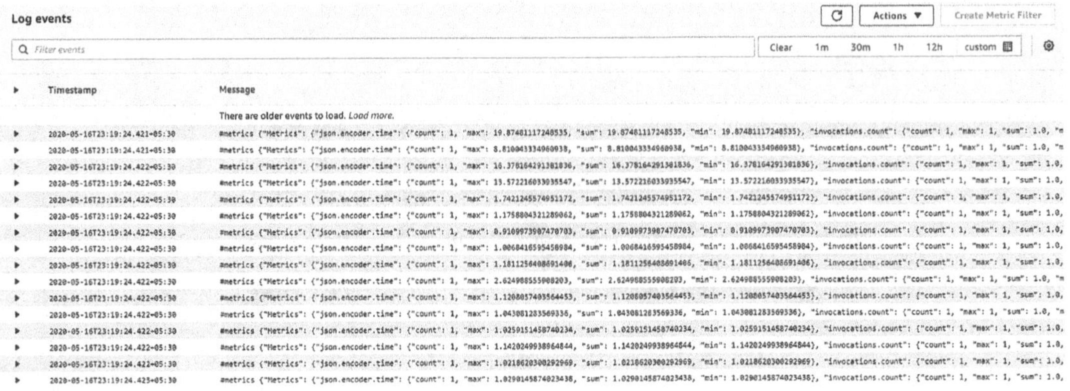

Figure 7-10. *Exploring the linear learning endpoint*

You can see all the test data that we sent for the predictions, and it is giving us JSON format output for that. This is how we can use CloudWatch Logs for our jobs. One more section under Logs is Transform Jobs, which we will look at in the next chapter once we discuss the batch transform job.

CloudWatch Metrics

Similar to how we can use CloudWatch Logs to view the logs of our jobs, we can get the metrics related to the algorithms or resources. Let's start with understanding the metrics related to the training jobs. We will log in to our SageMaker console and go to the training jobs page. We will explore the linear learner metrics for the classification task that we did on the Big Mart dataset. We will follow the same procedure to go to the algorithm page as we saw in CloudWatch Logs. Once on that page, instead of clicking to view the logs, we will click to view the metrics, as you can see in Figure 7-11.

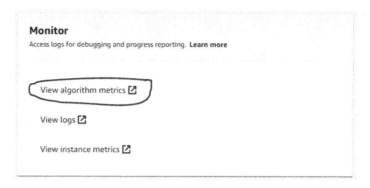

Figure 7-11. *Selecting metrics*

Once you click "View Algorithm Metrics," on the new screen you'll find different metrics available for that algorithm and a graph canvas. It will look like Figure 7-12.

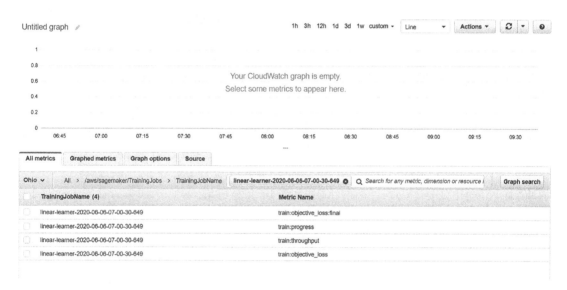

Figure 7-12. *Selecting the objective metrics*

In the linear learner algorithm, the metrics that we can see are the loss function, the training progress, and the throughput. Let's look at the loss function final value once the training stopped. We can click the "objective_loss" and then click "Add to graph." Next we can move to the "Graphed metrics" option and change the period to the time when you finished the training. You can change the graph type to Number. You will get the output, as shown in Figure 7-13.

Figure 7-13. *Visualizing objective loss*

Similarly, based upon the dataset, we can access a lot of metrics, as shown in Figure 7-14.

train:objective_loss	#quality_metric: host=\S+, epoch=\S+, train \S+_objective <loss>=(\S+)
validation:macro_precision	#quality_metric: host=\S+, validation macro_precision <score>=(\S+)
validation:dcg	#quality_metric: host=\S+, validation dcg <score>=(\S+)
test:mse	#quality_metric: host=\S+, test mse <loss>=(\S+)
validation:binary_f_beta	#quality_metric: host=\S+, validation binary_f_\S+ <score>=(\S+)
validation:objective_loss	#quality_metric: host=\S+, epoch=\S+, validation \S+_objective <loss>=(\S+)
validation:objective_loss:final	#quality_metric: host=\S+, validation \S+_objective <loss>=(\S+)
test:macro_recall	#quality_metric: host=\S+, test macro_recall <score>=(\S+)
test:absolute_loss	#quality_metric: host=\S+, test absolute_loss <loss>=(\S+)
train:recall	#quality_metric: host=\S+, train recall <score>=(\S+)
train:mse	#quality_metric: host=\S+, train mse <loss>=(\S+)
train:precision	#quality_metric: host=\S+, train precision <score>=(\S+)
train:objective_loss:final	#quality_metric: host=\S+, train \S+_objective <loss>=(\S+)
validation:recall	#quality_metric: host=\S+, validation recall <score>=(\S+)
test:multiclass_accuracy	#quality_metric: host=\S+, test multiclass_accuracy <score>=(\S+)
validation:precision	#quality_metric: host=\S+, validation precision <score>=(\S+)
validation:multiclass_accuracy	#quality_metric: host=\S+, validation multiclass_accuracy <score>=(\S+)
train:binary_f_beta	#quality_metric: host=\S+, train binary_f_\S+ <score>=(\S+)
test:recall	#quality_metric: host=\S+, test recall <score>=(\S+)
test:macro_precision	#quality_metric: host=\S+, test macro_precision <score>=(\S+)
test:macro_f_beta	#quality_metric: host=\S+, test macro_f_\S+ <score>=(\S+)
test:objective_loss	#quality_metric: host=\S+, test \S+_objective <loss>=(\S+)
test:precision	#quality_metric: host=\S+, test precision <score>=(\S+)
validation:multiclass_top_k_accuracy	#quality_metric: host=\S+, validation multiclass_top_k_accuracy_\S+ <score>=(\S+)
train:binary_classification_accuracy	#quality_metric: host=\S+, train binary_classification_accuracy <score>=(\S+)
validation:mse	#quality_metric: host=\S+, validation mse <loss>=(\S+)
test:multiclass_top_k_accuracy	#quality_metric: host=\S+, test multiclass_top_k_accuracy_\S+ <score>=(\S+)
validation:binary_classification_accuracy	#quality_metric: host=\S+, validation binary_classification_accuracy <score>=(\S+)
train:absolute_loss	#quality_metric: host=\S+, train absolute_loss <loss>=(\S+)
validation:macro_recall	#quality_metric: host=\S+, validation macro_recall <score>=(\S+)
train:throughput	#throughput_metric: host=\S+, train throughput=(\S+) records/second
test:binary_classification_accuracy	#quality_metric: host=\S+, test binary_classification_accuracy <score>=(\S+)
validation:absolute_loss	#quality_metric: host=\S+, validation absolute_loss <loss>=(\S+)
validation:macro_f_beta	#quality_metric: host=\S+, validation macro_f_\S+ <score>=(\S+)

Figure 7-14. *Different kinds of metrics available*

The same procedure can be adopted to get the metrics for any algorithm that has been run. Let's look at some aspects of instance metrics as well. In the instance metrics, we can look at the CPU utilization, memory utilization, and disk utilization, as shown in Figure 7-15.

Figure 7-15. *Visualizing metrics*

You can see the metrics changing live if you open CloudWatch during the training session.

Conclusion

In this chapter, you learned how to use CloudWatch to get the logs and metrics of different algorithms and resources. In the next chapter, we will look at how we can train a custom algorithm and also look at some of the other aspects of SageMaker.

CHAPTER 8

Running a Custom Algorithm in SageMaker

In this chapter, you will see how to run an algorithm of your own, instead of using SageMaker's built-in algorithms. Although SageMaker provides built-in algorithms for almost any kind of problem statement, many times we want to run our own custom model utilizing the power of SageMaker. We can do so effectively if we have working knowledge of Docker and hands-on knowledge of Python. In this chapter, we will create a custom random forest model for our Big Mart dataset. We will deploy the container in ECR and then train the model using SageMaker. Then we will use the model to do real-time inference as well as understand how batch transformation is done.

The Problem Statement

The problem statement is that we will try to predict the sales of an e-commerce firm using the random forest algorithm (one of the supervised learning ensemble tree algorithms). As it is a regression problem, we will be using the `RandomForestRegressor` class of the Scikit-Learn package. We have already explored the dataset in the previous chapters; it's the Big Mart dataset. Figure 8-1 shows the first few rows of the dataset.

© Himanshu Singh 2021
H. Singh, *Practical Machine Learning with AWS*, https://doi.org/10.1007/978-1-4842-6222-1_8

	Item_Identifier	Item_Weight	Item_Fat_Content	Item_Visibility	Item_Type	Item_MRP	Outlet_Identifier	Outlet_Establishment_Year	Outlet_Size
0	FDA15	9.30	Low Fat	0.016047	Dairy	249.8092	OUT049	1999	Medium
1	DRC01	5.92	Regular	0.019278	Soft Drinks	48.2692	OUT018	2009	Medium
2	FDN15	17.50	Low Fat	0.016760	Meat	141.6180	OUT049	1999	Medium
3	FDX07	19.20	Regular	0.000000	Fruits and Vegetables	182.0950	OUT010	1998	NaN
4	NCD19	8.93	Low Fat	0.000000	Household	53.8614	OUT013	1987	High

Figure 8-1. *Start of the dataset*

Review Chapter 5 if you want to understand the dataset. We already processed the data in that chapter and created the training and validate dataset, so we will not be repeating that process here. We will continue developing the algorithm after the train and test split is done. Please go through Chapter 5 to revise the entire process.

Running the Model

Before moving to the application of the model inside the SageMaker environment, let's first run the algorithm, locally, on the dataset that we have prepared and check the total loss that was incurred.

```
from sklearn.ensemble import RandomForestRegressor
rfc = RandomForestRegressor(n_estimators=500)
```

In the previous code, we initialized the RandomForestRegressor algorithm and asked to merge the outputs of 500 individual decision trees. Once we have initialized the algorithm, we can start training the model.

```
rfc.fit(X_train, y_train)
```

The previous code will start the training of the model. Now we can use the trained model to make predictions on the test set.

```
predictions = rfc.predict(X_test)
```

All the predictions are not stored in the variable predictions. Let's calculate the roto mean squared error of the model that we have created.

```
from sklearn.metrics import mean_squared_error
np.sqrt(mean_squared_error(predictions, y_test))
```

This will give a value for the final error. In my case, it's 1054 approximately. Different systems may have different outputs due to sampling.

Transforming Code to Use SageMaker Resources

Now that we have successfully run the code in the local environment, we will next transform it so that it can be run inside the SageMaker environment. The following are the steps to run a custom model in SageMaker:

1. Store the data in S3.

2. Create a training script and name it `train`.

3. Create an inference script that will help in predictions. We will call it `predictor.py`.

4. Set up files so that it will help in endpoint generation.

5. Create a Dockerfile that will help in building an image inside which the entire code will run.

6. Build a script to push the Docker image to Amazon Elastic Container Registry (ECR).

7. Use the SageMaker and Boto3 APIs to train and test the model.

We already have our training data inside S3, so we will start by creating a training script.

Creating the Training Script

We have already created a training notebook. This training script will be similar to the notebook, but we have a few extra considerations. The first thing that should be kept in mind is that the script is going to run inside a container. So, there can be a synchronization issue as the script is inside while the data is coming from S3 bucket, which is outside the container. Also, the results of the algorithm should also be saved in the S3 bucket. We need to keep all this in mind as we create a training script.

The first thing that we should know is that inside the container, no matter what the data is that is coming in, it gets stored inside the folder `/opt/ml`. Therefore, data from

S3 will be downloaded from that folder. So, in this folder we have to create three folders: one to store the input, one to store the output, and one to store the models. This can be defined by using the following script:

```
prefix = '/opt/ml/'
input_path = prefix + 'input/data'
output_path = os.path.join(prefix, 'output')
model_path = os.path.join(prefix, 'model')
```

Inside the data folder, we can have multiple files such as training, validation, or testing. We can also have separate files contributing to a single training file. Hence, we can make this kind of segregation as well. For us, we have only one file: the training file. So, we will be using only one channel.

```
channel_name='training'
training_path = os.path.join(input_path, channel_name)
```

This prepares our training script to handle data. Next is the training script itself. The data will come from S3. First we have to read it and then apply all the steps that we saw in Chapter 5. To read the file, we can use the following script:

```
input_files = [ os.path.join(training_path, file) for file in
os.listdir(training_path) ]
raw_data = [ pd.read_csv(file) for file in input_files ]
data = pd.concat(raw_data)
```

This script also helps if you have multiple CSV sheets to read. But, in that case remember to keep the parameter header=None. Now that we have read the data, we can start the training process. The following is the entire script for the training:

```
def train():
    print('Starting the training.')
    try:

        # Take the set of files and read them all into a single pandas
        dataframe
        input_files = [ os.path.join(training_path, file) for file in
        os.listdir(training_path) ]
```

```
if len(input_files) == 0:
raise ValueError(('There are no files in {}.\n' +
        'This usually indicates that the channel ({}) was
        incorrectly specified,\n' +
        'the data specification in S3 was incorrectly specified or
        the role specified\n' +
        'does not have permission to access the data.').
        format(training_path, channel_name))
raw_data = [ pd.read_csv(file) for file in input_files ]
data = pd.concat(raw_data)
data = data.sample(frac=1)

for i in data.Item_Type.value_counts().index:
    data.loc[(data['Item_Weight'].isna()) & (data['Item_Type'] == i),
    ['Item_Weight']] = \
    data.loc[data['Item_Type'] == 'Fruits and Vegetables',
    ['Item_Weight']].mean()[0]

cat_data = data.select_dtypes(object)
num_data = data.select_dtypes(np.number)

cat_data.loc[(cat_data['Outlet_Size'].isna()) & (cat_data['Outlet_
Type'] == 'Grocery Store'), ['Outlet_Size']] = 'Small'
cat_data.loc[(cat_data['Outlet_Size'].isna()) & (cat_data['Outlet_
Type'] == 'Supermarket Type1'), ['Outlet_Size']] = 'Small'
cat_data.loc[(cat_data['Outlet_Size'].isna()) & (cat_data['Outlet_
Type'] == 'Supermarket Type2'), ['Outlet_Size']] = 'Medium'
cat_data.loc[(cat_data['Outlet_Size'].isna()) & (cat_data['Outlet_
Type'] == 'Supermarket Type3'), ['Outlet_Size']] = 'Medium'

cat_data.loc[cat_data['Item_Fat_Content'] == 'LF' , ['Item_Fat_
Content']] = 'Low Fat'
cat_data.loc[cat_data['Item_Fat_Content'] == 'reg' , ['Item_Fat_
Content']] = 'Regular'
cat_data.loc[cat_data['Item_Fat_Content'] == 'low fat' , ['Item_
Fat_Content']] = 'Low Fat'
```

```
    le = LabelEncoder()
    cat_data = cat_data.apply(le.fit_transform)

    ss = StandardScaler()

    num_data = pd.DataFrame(ss.fit_transform(num_data.drop(['Item_
    Outlet_Sales'], axis=1)), columns = num_data.drop(['Item_Outlet_
    Sales'],axis=1).columns)
    cat_data = pd.DataFrame(ss.fit_transform(cat_data.drop(['Item_
    Identifier'], axis=1)), columns = cat_data.drop(['Item_
    Identifier'], axis=1).columns)

    final_data = pd.concat([num_data,cat_data],axis=1)

    X = final_data
    y = data['Item_Outlet_Sales']

    from sklearn.model_selection import train_test_split
    X_train, X_test, y_train, y_test = train_test_split(X, y,
    test_size = 0.1, random_state=5)

    from sklearn.ensemble import RandomForestRegressor
    rfc = RandomForestRegressor(n_estimators=500)

    clf = rfc.fit(X_train, y_train)

    # save the model
    with open(os.path.join(model_path, 'randomForest-tree-model.pkl'),
    'w') as out:
        pickle.dump(clf, out)
    print('Training complete.')
except Exception as e:
    trc = traceback.format_exc()
    with open(os.path.join(output_path, 'failure'), 'w') as s:
        s.write('Exception during training: ' + str(e) + '\n' + trc)
    print('Exception during training: ' + str(e) + '\n' + trc,
    file=sys.stderr)
    sys.exit(255)
```

We will keep the entire script inside a function called `train()`. After reading the CSV sheet, we will follow the same procedure we saw in Chapter 5. Later we will fit the random forest model on the data, which we ran in the previous section.

After all this, we have to save this model because later we will have to make predictions using the model. To save the model, we will first serialize it using pickle and then save it in the model location. Later, this model will be saved in S3.

Finally, we can run the entire script.

```
if __name__ == '__main__':
    train()
    sys.exit(0)
```

We have to use `sys.exit(0)` as it sends the message to SageMaker that the training has successfully completed. Save the file with the name `train` and no extension.

Creating the Inference Script

The training script is used to train the model. But, once the model is trained, we need to make predictions, whether with real-time inference as we saw in Chapter 6 or with the batch transformation that we will see in this chapter. We will save the inference script in a file named `predictor.py`.

The predictor file consists of the following components:

- `ScoringService()` class
- `ping()` method
- `transformation()` method
- Any other helper function required

The `ScoringService()` class consists of two functions. The first function, `get_model()`, loads and deserializes the model, while the second method, `predict()`, is responsible for making the predictions. Remember, the inference script also uses the same folder as the base that the training script uses, `/opt/ml`. The following is the script for the `ScoringService()` class:

```
prefix = '/opt/ml/'
model_path = os.path.join(prefix, 'model')
```

```
class ScoringService(object):
    model = None

    @classmethod
    def get_model(cls):
      if cls.model == None:
          with open(os.path.join(model_path, 'randomForest-tree-model.
          pkl'), 'r') as inp:
              cls.model = pickle.load(inp)
        return cls.model

    @classmethod
    def predict(cls, input):
        clf = cls.get_model()
        return clf.predict(input)
```

The ping() method is just used to check whether the Docker container that the code is running in is healthy. If it's not healthy, then it gives a 404 error, else 202.

```
@app.route('/ping', methods=['GET'])
def ping():
    status = 200 if health else 404
    return flask.Response(response='\n', status=status,
    mimetype='application/json')
```

transformation() is the method that is responsible for reading the test file and calling the required methods and classes. One thing to understand here is that this entire endpoint generation process is nothing but the creation of an API. Once the API is created, the data is sent as a POST request, and then we get the predictions as a response. This entire architecture is built using the Flask framework.

The data is sent using the POST method, so to read it, we need the StringIO() method to decode the data. Once the data is decoded, we can read it with our normal Pandas method. The transformation() function sends the data to the predict() function of class ScoringService(). The method sends the output back to the transformation() function. This prediction output is sent back to the host from where the API is called, with help from the StringIO() function. This finishes the entire cycle of endpoints. The following is the code of transformation():

```python
@app.route('/invocations', methods=['POST'])
def transformation():
    data = None

    if flask.request.content_type == 'text/csv':
        data = flask.request.data.decode('utf-8')
        s = StringIO.StringIO(data)
        data = pd.read_csv(s, header=None)
    else:
        return flask.Response(response='This predictor only supports CSV
        data', status=415, mimetype='text/plain')

    print('Invoked with {} records'.format(data.shape[0]))

    # Do the prediction
    predictions = ScoringService.predict(data)

    # Convert from numpy back to CSV
    out = StringIO.StringIO()
    pd.DataFrame({'results':predictions}).to_csv(out, header=False,
    index=False)
    result = out.getvalue()

    return flask.Response(response=result, status=200, mimetype='text/csv')
```

We will use this Python file for making the predictions, but to run the server efficiently, we need some configuration files. Let's explore them in the next section.

Configuring the Endpoint Generation Files

To run the inference server successfully, we need to configure the following files:

- nginx.conf file
- serve file
- wsgi.py file

Generally we don't make changes in these files. We create them and then use them as is for our predictions. We will not go into the line-by-line details of these files, but let's understand the purpose of each one.

The Nginx file is used to spin up the server and make the connection between the Docker containers deployed on EC2 instances and the client outside or inside the SageMaker network possible. Nginx uses a Python framework called Gunicorn that helps to set up the HTTP server.

Serve uses the running Gunicorn server to make the connection between the different resources feasible. Specifically, it is used for the following purposes:

- Efficiently using the number of CPUs for running the model

- Defining the server timeout

- Generating logs

- Starting the server using Nginx and Gunicorn

- Stopping the server if something doesn't go as expected

Lastly, the wsgi.py file is used to tell the server about our predictor.py file. We can explore the code in each file in the GitHub repository of this book. Remember, these files are really important as without them the server will never run; hence, you won't be able to make predictions. Don't make changes to these files, unless you are pretty sure about what you're doing.

Setting Up the Dockerfile

Now that all our script files are ready, we have to create a Docker image so that it can be uploaded to ECR and then SageMaker can access the code present in it and run it in an EC2 instance attached. Let's first see how to give a structure to the files that we created. Figure 8-2 depicts the structure that we should give to the directory, before creating the image.

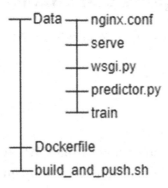

Figure 8-2. *Directory structure*

We have already created all the files that are present in the Data directory. Now, we have to create a Dockerfile script, which will be run to build the image. Then we will use the build_and_push.sh file to push the image to ECR.

These are the steps that we will follow in the Dockerfile:

1. Download an image from DockerHub that will have our operating system. We will download a minimal version of Ubuntu so that our code can run inside it. For this, we will use the following script:

   ```
   FROM ubuntu:16.04
   ```

2. Name the person, or the organization, who is maintaining and creating this image. I have given my name here. You can use any name you'd like.

   ```
   MAINTAINER Himanshu Singh
   ```

3. Run some Ubuntu commands so that we can set up the Python environment and update the operating system files. We will also download the server files that will be used to run the inference endpoints. You must be familiar with Linux commands to understand the script.

   ```
   RUN apt-get -y update && apt-get install -y --no-install-
   recommends \
           wget \
           python \
           nginx \
           ca-certificates \
       && rm -rf /var/lib/apt/lists/*
   ```

4. Once the setup is done, we can use pip from Python to install the important Python packages.

   ```
   RUN wget https://bootstrap.pypa.io/get-pip.py && python
   get-pip.py && \
       pip install numpy==1.16.2 scipy==1.2.1 scikit-learn==0.20.2
       pandas flask gevent gunicorn && \
           (cd /usr/local/lib/python2.7/dist-packages/scipy/.libs;
           rm *; ln ../../numpy/.libs/* .) && \
           rm -rf /root/.cache
   ```

5. Set the environment variables so that Python knows what the default folder is that will contain the code. Also, we will set some features of Python. We first make sure that timely log messages should be received from the container, and then we make sure that once any module is imported in Python, its `.pyc` file is not created. This is done using the variables `pythonunbuffered` and `pythondontwritebytecode`, respectively.

```
ENV PYTHONUNBUFFERED=TRUE
ENV PYTHONDONTWRITEBYTECODE=TRUE
ENV PATH="/opt/program:${PATH}"
```

6. Finally, the instance will instruct to copy our `Data` directory files to the default work directory, and then we will change the default work directory.

```
COPY Data /opt/program
WORKDIR /opt/program
```

This finishes our `Dockerfile` creation. Here is the entire code:

```
FROM ubuntu:16.04

MAINTAINER Himanshu Singh

RUN apt-get -y update && apt-get install -y --no-install-recommends \
        wget \
        python \
        nginx \
        ca-certificates \
    && rm -rf /var/lib/apt/lists/*

RUN wget https://bootstrap.pypa.io/get-pip.py && python get-pip.py && \
    pip install numpy==1.16.2 scipy==1.2.1 scikit-learn==0.20.2 pandas
    flask gevent gunicorn && \
        (cd /usr/local/lib/python2.7/dist-packages/scipy/.libs; rm *;
        ln ../../numpy/.libs/* .) && \
        rm -rf /root/.cache
```

```
ENV PYTHONUNBUFFERED=TRUE
ENV PYTHONDONTWRITEBYTECODE=TRUE
ENV PATH="/opt/program:${PATH}"

COPY Data /opt/program
WORKDIR /opt/program
```

Now, let's look at the script that we will use to push this image to ECR.

Pushing the Docker Image to ECR

We will create a shell script file, which will be used first to build the image from the Dockerfile that we created in the previous section and then to push the image to ECR. Let's look at the step-by-step procedure for this:

1. Name the image. We will save the name in a variable.

   ```
   algorithm_name=sagemaker-random-forest
   ```

2. Give full read and write permission to the train and serve files so that once the container is started, there are no access denied errors.

   ```
   chmod +x Data/train
   chmod +x Data/serve
   ```

3. Get AWS configurations so that there is no stoppage when the image is being pushed. We will define the account and the region of our AWS. Remember, since we will be running this code from inside SageMaker, the information can be automatically fetched. If we are running this from your local system or anywhere outside of AWS, then we will have to give the custom values.

   ```
   account=$(aws sts get-caller-identity --query Account
   --output text)
   region=$(aws configure get region)
   region=${region:-us-east-2}
   ```

4. Give the path and name to the container. We will use the same name that was given in the first step. We will use this path later to push the image.

```
fullname="${account}.dkr.ecr.${region}.amazonaws.com/${algorithm_
name}:latest"
```

5. Check whether the image already exists. If it doesn't, then a new image will be created; otherwise, the same image will be updated.

```
aws ecr describe-repositories --repository-names "${algorithm_
name}" > /dev/null 2>&1

if [ $? -ne 0 ]
then
    aws ecr create-repository --repository-name "${algorithm_
    name}" > /dev/null
fi
```

6. Get the login credentials of the AWS account.

```
$(aws ecr get-login --region ${region} --no-include-email)
```

7. Build the image with the name already decided, rename it with the full name we decided on that contains the ECR address, and then finally push the code.

```
docker build  -t ${algorithm_name} .
docker tag ${algorithm_name} ${fullname}
docker push ${fullname}
```

The following is the entire script that should be saved in the file build_and_push.sh.

```
algorithm_name=sagemaker-random-forest
chmod +x Container/train
chmod +x Container/serve
account=$(aws sts get-caller-identity --query Account --output text)
region=$(aws configure get region)
region=${region:-us-east-2}
fullname="${account}.dkr.ecr.${region}.amazonaws.com/${algorithm_name}:latest"
```

```
aws ecr describe-repositories --repository-names "${algorithm_name}" >
/dev/null 2>&1
if [ $? -ne 0 ]
then
    aws ecr create-repository --repository-name "${algorithm_name}" >
    /dev/null
fi
$(aws ecr get-login --region ${region} --no-include-email)
docker build  -t ${algorithm_name} .
docker tag ${algorithm_name} ${fullname}
docker push ${fullname}
```

Once this step is done, we have to go to the terminal, go inside the directory where your Dockerfile is present, and then type the following:

```
sh build_and_push.sh
```

This will start running the script and will successfully upload the image to ECR. You can then go to ECR and check whether the image exists. Figure 8-3 shows our image in ECR.

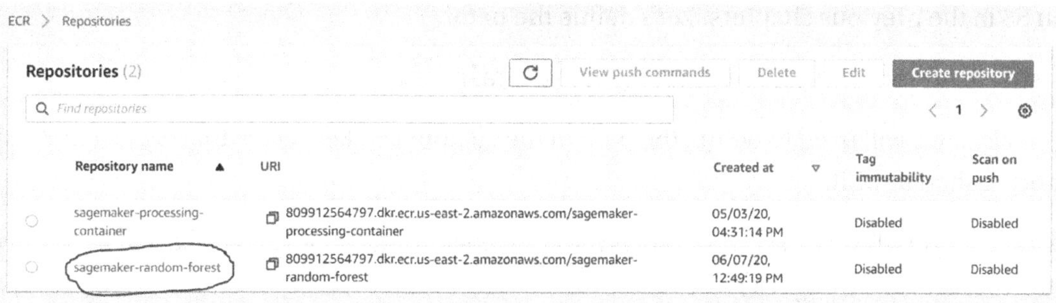

Figure 8-3. *Container image in ECR*

This finishes the process of creating the Docker . Now we will see in the next section how we can use this image to train the model in a SageMaker notebook.

Training the Model

Now we will use a SageMaker notebook to execute a classification model on the Big Mart dataset, using the random forest container we just created. Like the models that we saw in Chapter 6, we have to first define the role, then create a SageMaker session, and finally define the account and the region in which SageMaker is running.

```
import boto3
import re
import os
import numpy as np
import pandas as pd
from sagemaker import get_execution_role
import sagemaker as sage

role = get_execution_role()
sess = sage.Session()
account = sess.boto_session.client('sts').get_caller_identity()['Account']
region = sess.boto_session.region_name
```

The next step will be to get our data location. We have already uploaded our dataset in S3 in the previous chapters. Let's define the path.

```
data_location = 's3://slytherins-test/Train.csv'
```

Next, we will need to define the path of the custom Docker image that we just created and pushed to ECR.

```
image = '{}.dkr.ecr.{}.amazonaws.com/sagemaker-random-forest:latest'.
format(account, region)
```

Now that all the initial steps are done, we can start training our model. We will first initialize the container with the EC2 instance, image, role, and output S3 path, and then we will fit the model.

```
tree = sage.estimator.Estimator(image,
            role, 1, 'ml.m4.xlarge',
            output_path="s3://{}/output".format("slytherins-test"),
            sagemaker_session=sess)

tree.fit(data_location)
```

This will start the training job, and then once the job is finished, this will tell you the billable time as well. Figure 8-4 shows the output of the training job.

```
2020-06-07 07:27:23 Starting - Starting the training job...
2020-06-07 07:27:25 Starting - Launching requested ML instances........
2020-06-07 07:28:57 Starting - Preparing the instances for training...
2020-06-07 07:29:44 Downloading - Downloading input data...
2020-06-07 07:29:56 Training - Downloading the training image..Starting the training.

2020-06-07 07:30:27 Training - Training image download completed. Training in progress.
2020-06-07 07:31:13 Uploading - Uploading generated training modelTraining complete.

2020-06-07 07:32:39 Completed - Training job completed
Training seconds: 175
Billable seconds: 175
```

Figure 8-4. *Output of the training job*

This finishes our training job. We can always look at the detailed logs of the algorithm using CloudWatch.

Deploying the Model

Now that we have successfully trained the model, we can deploy it using the following line of script:

```
from sagemaker.predictor import csv_serializer
predictor = tree.deploy(1, 'ml.m4.xlarge', serializer=csv_serializer)
```

It will take some time to spin up an instance, and then it will be time to start our inference.

Doing Real-Time Inference

Let's use the test dataset of the Big Mart dataset and make predictions using the live endpoint we just deployed the model on.

```
predictions = predictor.predict(test_data.values).decode('utf-8')
```

Remember, to make these predictions, we need to have an endpoint up and running. This means when the endpoint is running, we will have to pay Amazon. But, if we want to start the endpoint only to make predictions and then automatically delete the endpoint, then we can use the Batch Transformation service in SageMaker. Let's look at this in the next section.

Doing Batch Transformation

To do the batch transformation, the first thing we need to do is to create a model that contains the model files that were generated when we trained the model and the image of the algorithm. The following is the code using that we can use to achieve this:

```
import boto3
client = boto3.client('sagemaker')
image = '{}.dkr.ecr.{}.amazonaws.com/sagemaker-random-forest:latest'.
format(account, region)
role = get_execution_role()
primary_container = {
    'Image': image,
    'ModelDataUrl': 's3://sagemaker-us-east-2-809912564797/output/
    sagemaker-random-forest-2020-06-07-07-27-23-190/output/model.tar.gz'
}

create_model_response = client.create_model(
    ModelName = 'Random-Forest-BigMart',
    ExecutionRoleArn = role,
    PrimaryContainer = primary_container)
```

This will package everything, and then we can pass it to the batch transform script so that we can start the predictions. To start a batch transform, first we need to store the test file in S3. Once we have stored the file, then we need to provide its location and the location in S3 where the predictions will be saved. We must give a unique name to the job.

The following is the script that will be used to run the batch transform job:

```
import time
from time import gmtime, strftime

batch_job_name = 'RF-Batch-Transform-' + strftime("%Y-%m-%d-%H-%M-%S",
gmtime())
input_location = 's3://slytherins-test/test_data.csv'
output_location = 's3://{}/{}/output/{}'.format('slytherins-test',
'RF-Batch-Transform', batch_job_name)

request = \
{
    "TransformJobName": 'Random-Forest-BigMart-1',
    "ModelName": 'Random-Forest-BigMart',
    "TransformOutput": {
        "S3OutputPath": output_location,
        "Accept": "text/csv",
        "AssembleWith": "Line"
    },
    "TransformInput": {
        "DataSource": {
            "S3DataSource": {
                "S3DataType": "S3Prefix",
                "S3Uri": input_location
            }
        },
        "ContentType": "text/csv",
        "SplitType": "Line",
        "CompressionType": "None"
    },
    "TransformResources": {
            "InstanceType": "ml.m4.xlarge",
            "InstanceCount": 1
    }
}
```

```
client.create_transform_job(**request)
print("Created Transform job with name: ", batch_job_name)
```

Some of the keys used in the previous code are explained here:

- ContentType tells about the data type of our test dataset. It's a CSV in our case.

- SplitType tells how different rows are split in our dataset. It is split by line in our case.

- CompressionType tells whether our data is raw or it is a compressed file like a TAR file. For us it is a raw CSV file.

Once we execute the previous code, we can monitor the job progress in the SageMaker console. But, if we want to monitor the progress directly in the notebook, we can use the following script:

```
try: client.get_waiter('transform_job_completed_or_stopped').
wait(TransformJobName='Random-Forest-BigMart-1')
finally:
    response = client.describe_transform_job(TransformJobName='Random-
    Forest-BigMart-1')
    status = response['TransformJobStatus']
    print("Transform job ended with status: " + status)
    if status == 'Failed':
        message =response['FailureReason']
        print('Transform failed with the following error: {}'.
        format(message))
        raise Exception('Transform job failed')
```

To look at the job in the SageMaker console, we need to select the batch transform option. See Figure 8-5.

▼ Notebook

   Notebook instances

   Lifecycle configurations

   Git repositories

▼ Training

   Algorithms

   Training jobs

   Hyperparameter tuning jobs

▼ Inference

   Compilation jobs

   Model packages

   Models

   Endpoint configurations

   Endpoints

   Batch transform jobs

Figure 8-5. *Batch transform job*

You can use CloudWatch for getting the logs and the metrics. We can find out whether the job is completed successfully by looking at the status. See Figure 8-6.

Job summary

Job name	Status	Approx. batch transform duration
Random-Forest-BigMart-1	⊘ Completed	2 minute(s)
Job ARN	Creation time	
arn:aws:sagemaker:us-east-2:809912564797:transform-job/random-forest-bigmart-1	Jun 07, 2020 14:00 UTC	

Figure 8-6. *Status of the job*

Once the job is completed, we can go to the S3 output location and look at the predictions.

Conclusion

In this chapter, you learned how to create custom containers to run the code and algorithms that are not present in SageMaker, while using the computational power of AWS and services of SageMaker. We can run any kind of custom code by following the same procedure.

In the next chapter, you will learn how we can create an end-to-end pipeline using Step Functions.

CHAPTER 9

Making an End-to-End Pipeline in SageMaker

In this chapter, we will see how we can make an end-to-end pipeline of an entire machine learning process. We can use a combination of AWS services to automate the entire process of machine learning. All the processes that we have seen in the previous chapters, from the data processing steps to the endpoint generation, can be automated and then be run directly with a click of a button. The only thing we need to change is the dataset, but the process remains the same. Let's see how we can automate what we did on the Big Mart dataset in the previous chapters.

Let's start by looking at AWS Step Functions.

Overview of Step Functions

AWS Step Functions is the service provided by Amazon that you can use to create workflows and automate them. These workflows consist of AWS resources, algorithms, and processing. They may also include resources that are outside AWS. We can use Step Functions to create an end-to-end automation framework that helps us in building an effective continuous integration and continuous development (CI/CD) DevOps pipeline.

Each component in a step function is called a *state machine*. In this chapter, we will be creating multiple state machines, as follows:

- State machine for training a model
- State machine for saving the model
- State machine for configuring endpoints
- State machine for model deployment

© Himanshu Singh 2021
H. Singh, *Practical Machine Learning with AWS*, https://doi.org/10.1007/978-1-4842-6222-1_9

Then we will combine all the state machines in a sequential format so that the entire process can be automated. Figure 9-1 is a small workflow that shows how these state machines will be connected.

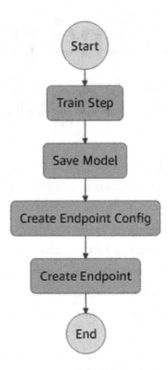

Figure 9-1. *Steps involved in creating a state machine*

Let's start the process of creating the previous workflow. The first step will be to upgrade the Step Functions package so that we can make sure we are using the latest version of the module.

Upgrading Step Functions

We will simply use pip from Python to upgrade the Step Functions package and all the dependent packages.

```
python -m pip install --upgrade stepfunctions
```

You can run this either from the terminal or from the Jupyter Notebook as well by adding a prefix of exclamation mark (!).

Defining the Required Parameters

Let's now define the required objects that we will use to run our code. This includes the roles, region, bucket, etc.

```
import boto3
import sagemaker
import time
import random
import uuid
import logging
import stepfunctions
import io
import random
from sagemaker.amazon.amazon_estimator import get_image_uri
from stepfunctions import steps
from stepfunctions.steps import TrainingStep, ModelStep, TransformStep
from stepfunctions.inputs import ExecutionInput
from stepfunctions.workflow import Workflow
from stepfunctions.template import TrainingPipeline
from stepfunctions.template.utils import replace_parameters_with_jsonpath

sagemaker_execution_role = sagemaker.get_execution_role()
workflow_execution_role = "arn:aws:iam::809912564797:role/himanshu-step-
functions"
session = sagemaker.Session()
stepfunctions.set_stream_logger(level=logging.INFO)
region = boto3.Session().region_name
prefix = 'sagemaker/big-mart-pipeline'
bucket_path = 'https://s3-{}.amazonaws.com/{}'.format(region, "slytherins-
test")
```

As you can see in the code, we require two roles. One is the SageMaker execution role, and the second is the workflow execution role. In the next section, we will see how to define the role for workflow execution. In addition, we have created a SageMaker session and defined the region and S3 bucket location. We have also set the Step Functions logger so that whatever important messages are there, we will not miss them.

Now let's see how we can create the required IAM role for workflow execution.

Setting Up the Required Roles

We need to set up two things to be able to execute the workflow:

1. We need to add a policy on the already existing SageMaker role.

2. We need to create a new Step Functions IAM role.

Adding a Policy to the Existing SageMaker Role

For the current SageMaker role that we are using to run all the models in this entire book, it's easy to update the policy so that it can access the features of Step Functions. In the SageMaker console, we need to click the name of the notebook instance that we are using. This will lead us to a page showing the properties of the notebook instance. In that page there will be a section named "Permissions and encryption." There you will find your ARN role mentioned for the instance. See Figure 9-2.

Figure 9-2. *Selecting the ARN role*

Once you click that role, you'll move to the IAM role for that ARN. On that page, you'll need to click Attach Policies and search for *AWSStepFunctionsFullAccess*. Attach this policy, and now your SageMaker instance is ready to use Step Functions. See Figure 9-3.

Figure 9-3. *Attaching policies*

Creating a New IAM Role for Step Functions

Once we are done with enabling the instance to execute a Step Functions job, we need to create an execution role so that Step Functions is able to execute the jobs that are created. For this, again we need to go to the IAM console and create this role.

Go to the IAM console, go to the Roles section, and then click "Create role." See Figure 9-4.

Figure 9-4. *Creating roles*

Select the Step Functions service. You may need to search for the service. See Figure 9-5.

Common use cases

EC2
Allows EC2 instances to call AWS services on your behalf.

Lambda
Allows Lambda functions to call AWS services on your behalf.

Or select a service to view its use cases

API Gateway	CodeGuru	ElastiCache	Kinesis	RoboMaker
AWS Backup	CodeStar Notifications	Elastic Beanstalk	Lake Formation	S3
AWS Chatbot	Comprehend	Elastic Container Service	Lambda	SMS
AWS Support	Config	Elastic Transcoder	Lex	SNS
Amplify	Connect	ElasticLoadBalancing	License Manager	SWF
AppStream 2.0	DMS	Forecast	Machine Learning	SageMaker
AppSync	Data Lifecycle Manager	GameLift	Macie	Security Hub
Application Auto Scaling	Data Pipeline	Global Accelerator	MediaConvert	Service Catalog
Application Discovery Service	DataSync	Glue	Migration Hub	Step Functions
	DeepLens	Greengrass	OpsWorks	Storage Gateway
Batch	Directory Service	GuardDuty	Personalize	Systems Manager
Chime	DynamoDB	Health Organizational View	Purchase Orders	Textract
CloudFormation	EC2	IAM Access Analyzer	QLDB	Transfer
CloudHSM	EC2 - Fleet	Inspector	RAM	Trusted Advisor
CloudTrail	EC2 Auto Scaling	IoT	RDS	VPC
CloudWatch Application Insights	EC2 Image Builder	IoT Things Graph	Redshift	WorkLink
CloudWatch Events	EKS	KMS	Rekognition	WorkMail
CodeBuild	EMR			
CodeDeploy				

Figure 9-5. *Selecting the Step Functions service*

Now, continue the process and keep clicking Next until you arrive at the section where you need to provide the role name. Give any role name you want and then click "Create role." Next, once we have created the role, we need to attach a policy to it. Here we will list all the services that the Step Functions service is allowed to do. We provide this list in JSON format.

Click the role that you have just created, and then in the Permissions section click "Add inline policy." See Figure 9-6.

Figure 9-6. *Adding inline policies*

Here, you need to add a JSON file on the JSON tab. The file contents are shown here:

```
{
  "Version": "2012-10-17",
  "Statement": [
      {
          "Effect": "Allow",
          "Action": [
            "sagemaker:CreateTransformJob",
            "sagemaker:DescribeTransformJob",
            "sagemaker:StopTransformJob",
            "sagemaker:CreateTrainingJob",
            "sagemaker:DescribeTrainingJob",
            "sagemaker:StopTrainingJob",
            "sagemaker:CreateHyperParameterTuningJob",
            "sagemaker:DescribeHyperParameterTuningJob",
            "sagemaker:StopHyperParameterTuningJob",
            "sagemaker:CreateModel",
            "sagemaker:CreateEndpointConfig",
            "sagemaker:CreateEndpoint",
            "sagemaker:DeleteEndpointConfig",
            "sagemaker:DeleteEndpoint",
```

```
            "sagemaker:UpdateEndpoint",
            "sagemaker:ListTags",
            "lambda:InvokeFunction",
            "sqs:SendMessage",
            "sns:Publish",
            "ecs:RunTask",
            "ecs:StopTask",
            "ecs:DescribeTasks",
            "dynamodb:GetItem",
            "dynamodb:PutItem",
            "dynamodb:UpdateItem",
            "dynamodb:DeleteItem",
            "batch:SubmitJob",
            "batch:DescribeJobs",
            "batch:TerminateJob",
            "glue:StartJobRun",
            "glue:GetJobRun",
            "glue:GetJobRuns",
            "glue:BatchStopJobRun"
            ],
        "Resource": "*"
},
{
        "Effect": "Allow",
        "Action": [
            "iam:PassRole"
    ],
        "Resource": "*",
        "Condition": {
          "StringEquals": {
              "iam:PassedToService": "sagemaker.amazonaws.com"
            }
        }
},
```

```
{
    "Effect": "Allow",
    "Action": [
        "events:PutTargets",
        "events:PutRule",
        "events:DescribeRule"
    ],
    "Resource": [
        "arn:aws:events:*:*:rule/StepFunctionsGetEventsForSageMaker
        TrainingJobsRule",
        "arn:aws:events:*:*:rule/StepFunctionsGetEventsForSageMaker
        TransformJobsRule",
        "arn:aws:events:*:*:rule/StepFunctionsGetEventsForSageMaker
        TuningJobsRule",
        "arn:aws:events:*:*:rule/StepFunctionsGetEventsForECSTaskRule",
        "arn:aws:events:*:*:rule/StepFunctionsGetEventsForBatchJobsRule"
    ]
  }
 ]
}
```

Once that's done, you can review the policy, give it a name, and then create the policy. Don't forget to copy the ARN number of the policy you just created. This will help you when creating code in SageMaker.

Setting Up the Training Step

In the previous section, we completed all the necessary configuration steps to run our code to create a pipeline. In this section, we will create the first step: TrainingStep. The first thing that we will do is to create a dictionary that will auto-initialize the training job name, the model name, and the endpoint name. We can do so using the following code:

```
names = {
    'JobName': str,
    'ModelName': str,
```

```
    'EndpointName': str
}

execution_input = ExecutionInput(schema=names)
```

Next, we will create a training step by using the XGBoost container that we already learned about in the previous chapters. The first step will be to initialize the container.

```
tree = sage.estimator.Estimator(image,
          sagemaker_execution_role, 1, 'ml.m4.xlarge',
          output_path="s3://{}/output".format("slytherins-test"),
          sagemaker_session=sess)
```

Next, we need to create the training step. This is done by providing the path to the input training and validation data.

```
training_step = steps.TrainingStep(
    'Train Step',
    estimator=tree,
    data={
        'train': sagemaker.s3_input("s3://slytherins-test/Train.csv",
        content_type='text/csv'),
        'validation': sagemaker.s3_input("s3://slytherins-test/test_data.
        csv", content_type='text/csv')
    },
    job_name=execution_input['JobName']
)
```

Remember, this will not execute the model. Only a step is created here. First, we will create all the steps and then combine them and run them sequentially. Let's now decide on the step for saving the model. Once in the pipeline, the previous training is finished, and the model artifacts that are generated should be saved. That is done using the following code:

```
model_step = steps.ModelStep(
    'Save model',
    model=training_step.get_expected_model(),
    model_name=execution_input['ModelName']
)
```

The next step after the model is created is to define the configuration of the endpoint. Let's see that in the next section.

Setting Up the Endpoint Configuration Step

In this step, we will define what kind of resources are required to deploy the endpoint. We have already seen how an endpoint is deployed, so the step that we will create here will be self-explanatory.

```
endpoint_config_step = steps.EndpointConfigStep(
    "Create Endpoint Config",
    endpoint_config_name=execution_input['ModelName'],
    model_name=execution_input['ModelName'],
    initial_instance_count=1,
    instance_type='ml.m4.xlarge'
)
```

Once our configuration is done, we will create the step that will actually deploy the endpoint. Let's see that in the next section.

Setting Up the Endpoint Step

The following code creates a step that is used for the endpoint deployment:

```
endpoint_step = steps.EndpointStep(
    "Create Endpoint",
    endpoint_name=execution_input['EndpointName'],
    endpoint_config_name=execution_input['ModelName']
)
```

Once the endpoint is deployed, we can start the inference as we saw in the previous sections. Now that we have successfully created all the steps, let's join them together and create a sequence in the next section.

Creating a Chain of the Steps

To create a chain, we will start with the training step, then move on to the model saving step, then configure the endpoint, and finally deploy the model on the endpoint configured. We can create this chain using the following code:

```
workflow_definition = steps.Chain([
    training_step,
    model_step,
    endpoint_config_step,
    endpoint_step
])
```

Defining the Workflow and Starting Operation

Now that the components are connected in the previous step, we need to provide all the necessary configurations so that this workflow can be executed. This can be done using the following code:

```
workflow = Workflow(
    name='Big-Mart_Workflow-v1',
    definition=workflow_definition,
    role=workflow_execution_role,
    execution_input=execution_input
)
```

Once this is done, all we need to do is execute the workflow created. This can be done by using the following code:

```
workflow.create()
execution = workflow.execute(
    inputs={
        'JobName': 'regression-{}'.format(uuid.uuid1().hex),
        'ModelName': 'regression-{}'.format(uuid.uuid1().hex),
        'EndpointName': 'regression-{}'.format(uuid.uuid1().hex)
    }
)
```

Now, as you execute the previous code, the entire pipeline starts running. To see how the pipeline looks, you can use the render_graph() function.

workflow.render_graph()

You will see the pipeline shown in Figure 9-7.

Figure 9-7. *Rendering the graph*

You can also check the current progress of the process executed, by using the render_progress() function. See Figure 9-8.

execution.render_progress()

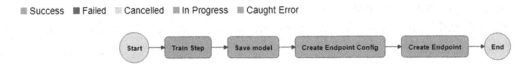

Figure 9-8. *Checking the progress*

As mentioned in the image, if any step is failed, it will be shown in red, otherwise green. We will explore more about this process by going to the Step Functions console. Let's see that in the next section.

Exploring the Jobs in Step Functions

Let's see how the workflow that we have created in this chapter looks in the Step Functions console. Search for *Step Functions* in the services list and click it. This will open your console. You can find the step function that has been created in this chapter, mentioned there. See Figure 9-9.

Figure 9-9. *Selecting the state machine*

Click the name of your pipeline and then click the latest job that ran or is running. See Figure 9-10.

Figure 9-10. *Selecting the latest job*

Here, you can see the pipeline that you have created. It will show all the steps beautifully in the dashboard. You can click the individual components and look at their progress as well. See Figure 9-11.

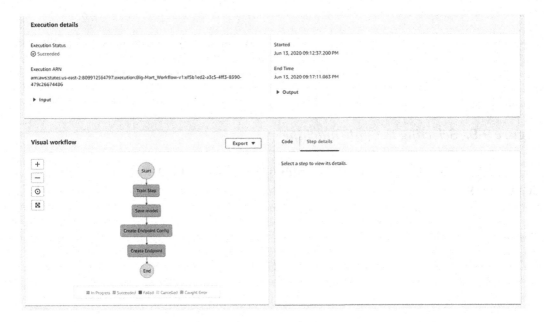

Figure 9-11. *Visualizing the progress*

If you click the train step, you can see its own process. See Figure 9-12.

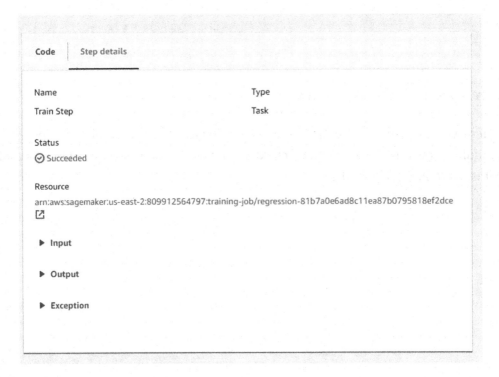

Figure 9-12. *Checking the progress state*

Exploring the JSON File That Can Be Passed as Input

There is one more option that we can use to create the flow in Step Functions if we want to avoid the Python code. You can pass a JSON directly and then decide on the sequence. Let's see the JSON that was generated by Step Functions for our code. To look at the JSON code, we have to click the Edit State Machine option. You can look at the entire JSON format, as shown in Figure 9-13.

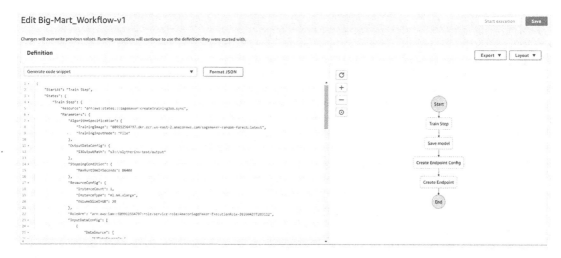

Figure 9-13.   *Looking at the JSON structure*

Conclusion

In this chapter, you learned how to create an end-to-end pipeline using Step Functions. This is useful when creating the entire training and deployment process and when retraining models with the new data or with some new configuration. This also helps in creating a CI/CD pipeline where we can push the code to Git and then use tools such as Jenkins or Bamboo to create these step functions and start the execution. Hence, as you push to code to Git, immediately the process of training starts. That's the power of creating a pipeline.

This finishes our discussion on SageMaker and its services. In the next chapter of this book, we will look at some of the use cases of machine learning that can be done using other AWS services.

PART III

Other AWS Services

## CHAPTER 10

Machine Learning Use Cases in AWS

In this chapter, we will explore three use cases in which AWS can be used to solve typical machine learning problems, without writing too much code. We will use some of the AWS services besides SageMaker.

Use Case 1: Natural Language Processing Using Amazon Comprehend

Amazon Comprehend is a service in AWS that can perform various NLP tasks such as key entities extraction from text, sentiment analysis, phrases extraction, syntax check, language detection, topic modeling, and document classification. Amazon Comprehend provides a UI that we can use to train a model without writing any code. We can also use the API provided by Comprehend to connect it to the scripting language of your choice and then train the model directly from the code.

In this section, we will explore how we can use Comprehend to analyze the text, and then we will create a custom sentiment analysis model using it.

Analysis of Text

From the Amazon Management Console, we can select the Amazon Comprehend service and then click Launch Amazon Comprehend. Then we will see a section called "Real-time analysis." Let's use this section to analyze some text in real time. The text that I have used for analysis is an excerpt from a *Game of Thrones* review: `https://www.polygon.com/tv/2019/6/3/18634311/game-of-thrones-review-full-tv-series-hbo`.

© Himanshu Singh 2021
H. Singh, *Practical Machine Learning with AWS*, https://doi.org/10.1007/978-1-4842-6222-1_10

The excerpt that we will analyze is as follows:

> "The series' ending unleashed a seemingly bottomless geyser
> of fan discontent ranging from mile-long Twitter threads to an
> honest-to-God petition for HBO to remake the eighth season
> from scratch. The complaints, by and large, feel typical to the
> "Peak TV" era: the uproar you'd expect from the sort of people
> who've interpreted Emilia Clarke's traumatized, brutal Daenerys
> Targaryen as a one-dimensional message about girl power; anger
> that such and such a character "deserved" some specific ending
> they didn't receive. Much of it boils down to viewers interpreting
> their own discomfort over the show's failures."

Figure 10-1 shows an "Input text" box. Just paste the previous excerpt there and click
Analyze.

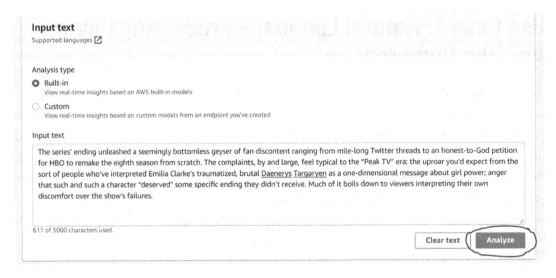

Figure 10-1. *Inputting text into Comprehend*

Now, let's see what Comprehend has given us after its analysis. The first thing is the
list of entities that it has extracted, along with the confidence level. See Figure 10-2.

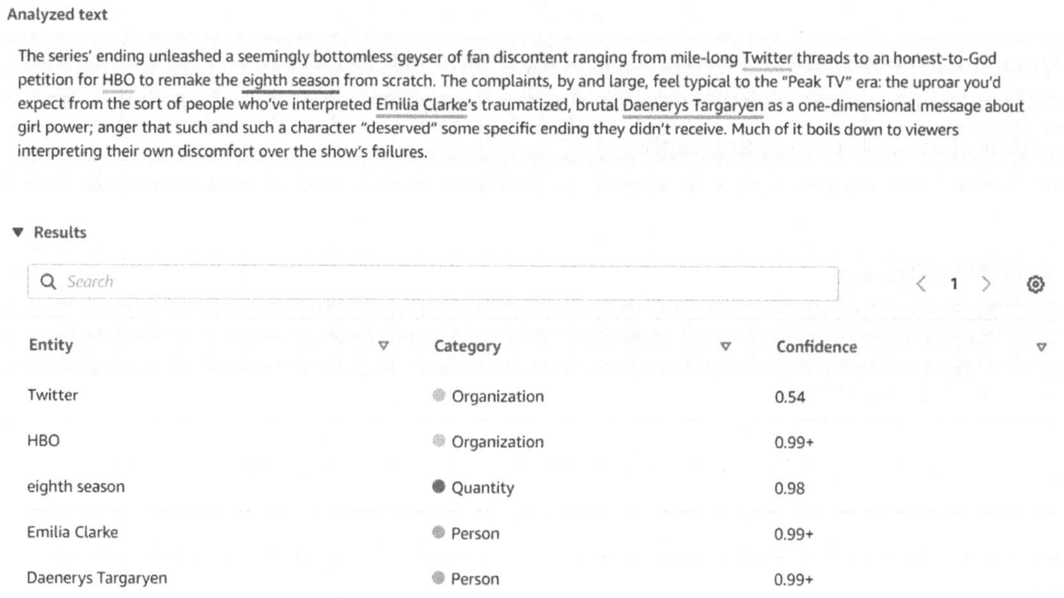

Analyzed text

The series' ending unleashed a seemingly bottomless geyser of fan discontent ranging from mile-long Twitter threads to an honest-to-God petition for HBO to remake the eighth season from scratch. The complaints, by and large, feel typical to the "Peak TV" era: the uproar you'd expect from the sort of people who've interpreted Emilia Clarke's traumatized, brutal Daenerys Targaryen as a one-dimensional message about girl power; anger that such and such a character "deserved" some specific ending they didn't receive. Much of it boils down to viewers interpreting their own discomfort over the show's failures.

▼ Results

Entity	Category	Confidence
Twitter	Organization	0.54
HBO	Organization	0.99+
eighth season	Quantity	0.98
Emilia Clarke	Person	0.99+
Daenerys Targaryen	Person	0.99+

Figure 10-2. *Result of analysis*

In Figure 10-2, you can see that most of the entities extracted are correct. Now let's see what the key phrases are in the text that Comprehend feels are important. Click the "Key phrases" tab. Figure 10-3 shows the output.

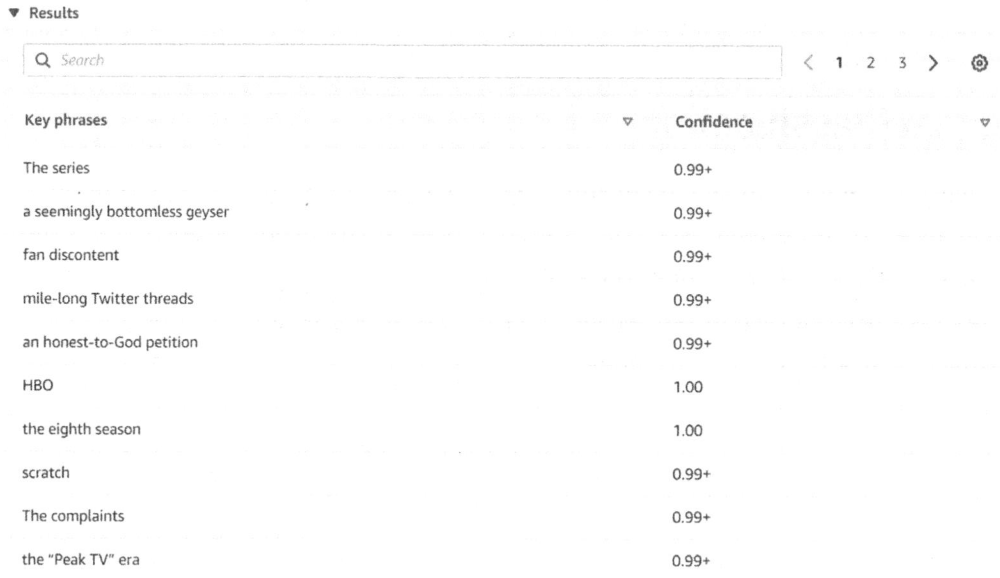

▼ Results

Key phrases	Confidence
The series	0.99+
a seemingly bottomless geyser	0.99+
fan discontent	0.99+
mile-long Twitter threads	0.99+
an honest-to-God petition	0.99+
HBO	1.00
the eighth season	1.00
scratch	0.99+
The complaints	0.99+
the "Peak TV" era	0.99+

Figure 10-3. *Key phrases present in the text*

Some of the phrases are really important, as by reading them we can tell what the paragraph is about, as well as its tone. Talking about the tone, let's look at the sentiment of the entire paragraph. For this we will click the Sentiment tab. Figure 10-4 shows the sentiment analysis of the paragraph.

▼ Results

Sentiment

Neutral	Positive	Negative	Mixed
0.59 confidence	0.17 confidence	0.23 confidence	0.00 confidence

Figure 10-4. *Sentiment analysis output*

Here you can see that Comprehend is telling us that the author is mostly neutral about *Game of Thrones*. If you read the paragraph, you'll see the author is actually telling about fans who didn't like the final season, rather than the author not liking it. Hence, in this section, Comprehend was able to give us the correct picture.

If you want to empower your scripts with Comprehend, you can use the API that it provides. It is beyond the scope of this chapter, but it's worth trying. You can read the Comprehend API documentation to try it.

Next, let's see how we can make a custom classification model using Comprehend. Here, we will be finding the toxicity level of a bunch of text. The dataset is taken from Kaggle, and you can download it from `www.kaggle.com/c/jigsaw-toxic-comment-classification-challenge/data`.

Custom Classification

The dataset contains a total of eight columns. I have kept only two columns and deleted the rest of them, as multiclass classification in Comprehend expects only two columns, one of the text and the other of the classes.

In the Comprehend console, we will click "Custom classification." Then click "Train classifier." Figure 10-5 shows the steps.

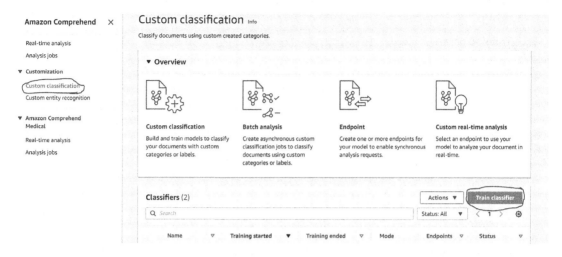

Figure 10-5. *Training a custom classifier*

The first step is to give your training job a name. Follow the instruction for the naming convention that Comprehend expects. See Figure 10-6.

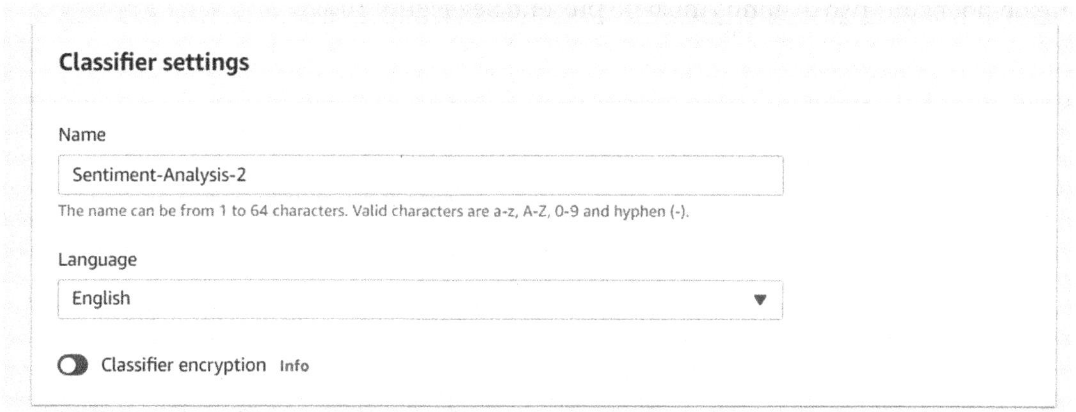

Figure 10-6. *Naming the classifier*

Next, you have to select a classification mode. Multiclass classification is where we have only one column of categories, while multilabel classification is where each class can have subcategories as well. We will select the first one, as shown in Figure 10-7.

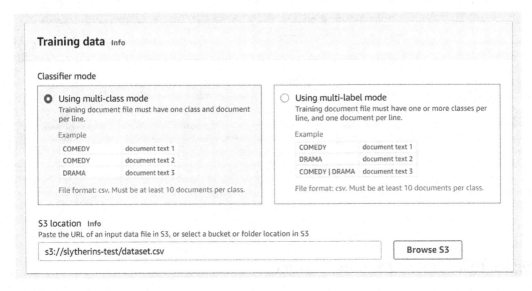

Figure 10-7. *Multiclass Comprehend classification*

Next, we have to upload the dataset to an S3 bucket and provide the path. Remember, only two columns must be present. See Figure 10-8.

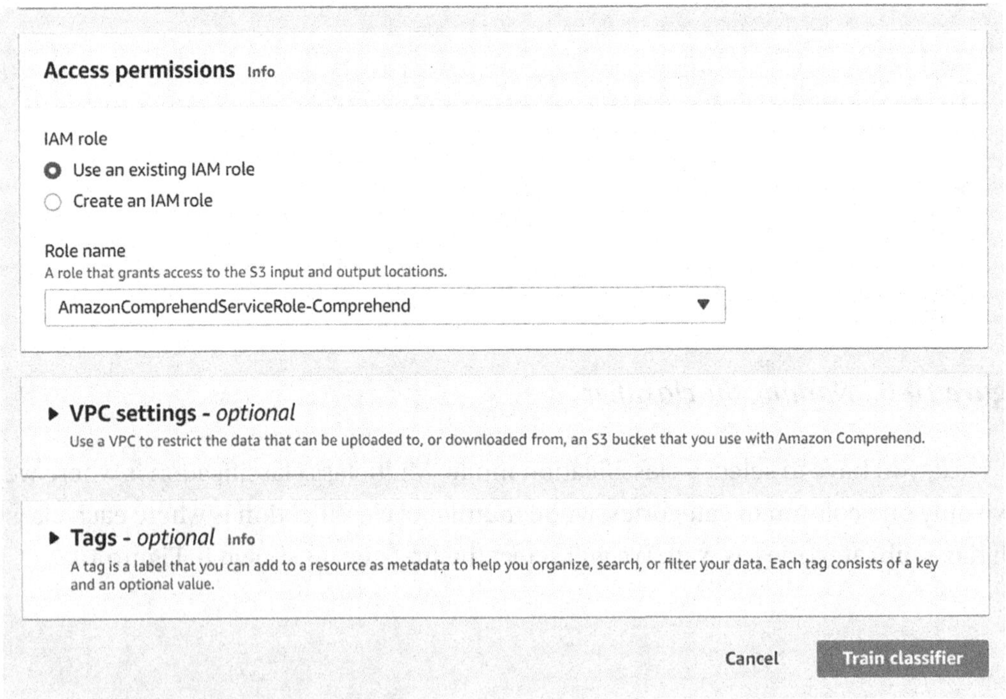

Figure 10-8. *Defining the IAM role*

Once you have provided the path, select your IAM role and then click "Train classifier." This will start the training of the model. Once your model is trained, you will see information similar to Figure 10-9.

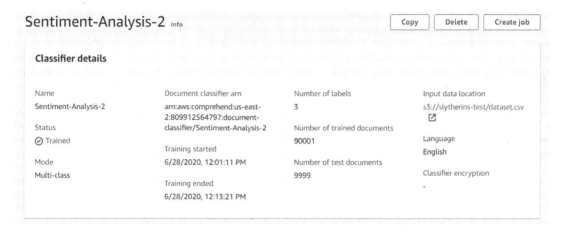

Figure 10-9. *Model training output*

You can next find the metrics of the trained model under "Classifier performance," as shown in Figure 10-10.

Classifier performance Info

Accuracy	Precision	Recall	F1 score
0.9575	0.9054	0.8356	0.86

Hamming loss	Micro precision	Micro recall	Micro F1 score
0.0425	0.9575	0.9575	0.9575

Figure 10-10. *Performance of the Comprehend classifier*

You can see that it gives a pretty good result, in terms of accuracy, precision, and recall. Next, to use this model, we can create an endpoint and then start the inference. Just click the "Create endpoint" button and give it a name to start the process. See Figure 10-11.

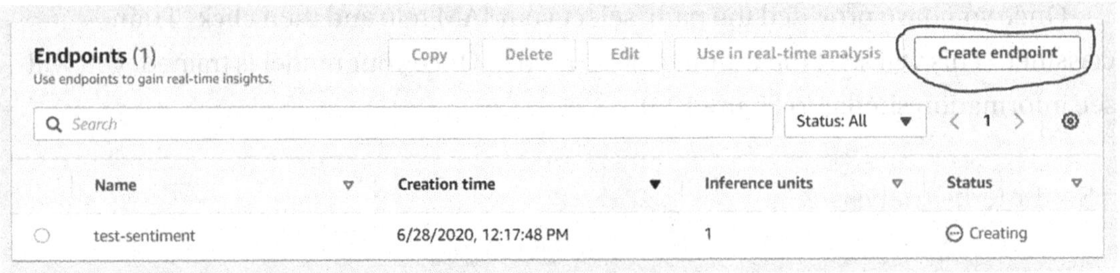

Figure 10-11. *Creation of endpoint*

Once the endpoint is ready, we can go to the real-time analysis section, and this time instead of clicking the built-in analysis type, click Custom and select your endpoint. Next, give any text and it will predict its class, as you can see in Figures 10-12 and 10-13.

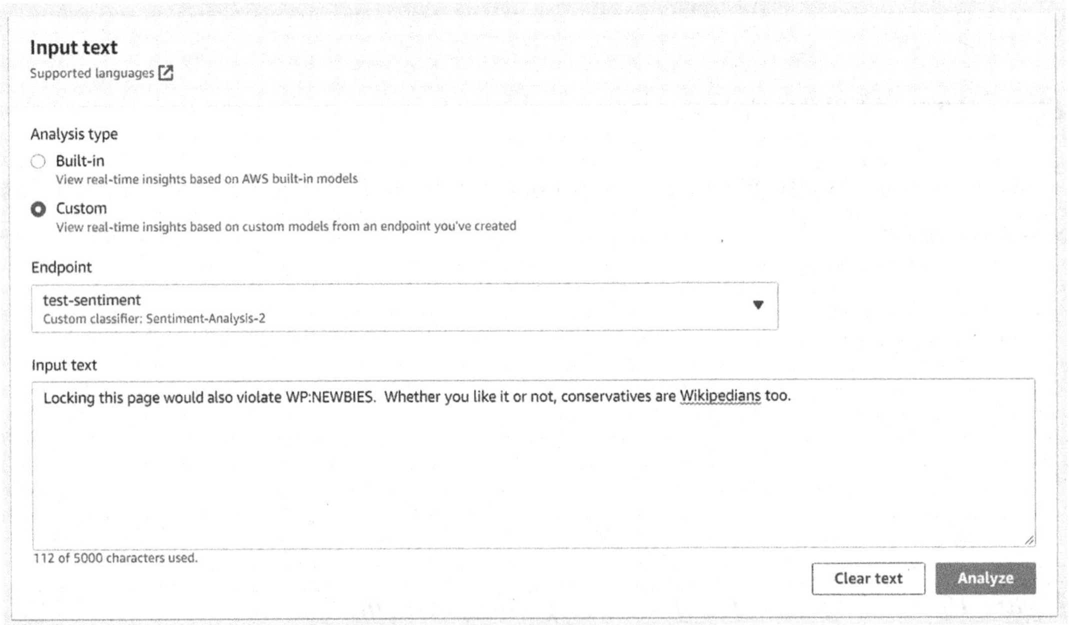

Figure 10-12. *Inference using the Custom classifier*

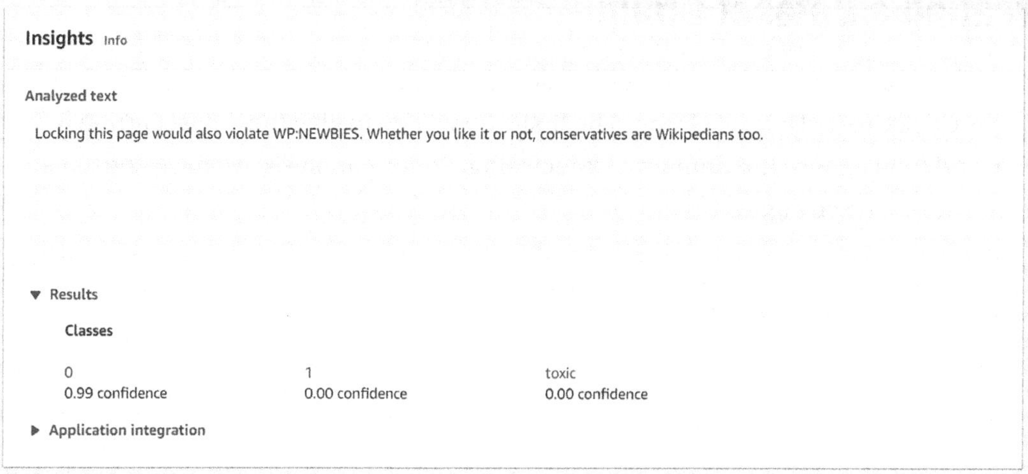

Insights Info

Analyzed text

Locking this page would also violate WP:NEWBIES. Whether you like it or not, conservatives are Wikipedians too.

▼ Results

Classes

0	1	toxic
0.99 confidence	0.00 confidence	0.00 confidence

▶ Application integration

Figure 10-13. Insights from the inference

Don't forget to delete the endpoint once the analysis is done, because it's chargeable. This finishes our discussion about the first use case. Next, let's look at a sales forecast model that can be built using another Amazon service called Amazon Forecast.

Use Case 2: Sales Forecasting Using Amazon Forecast

In this section, we will be predicting the sales forecast for a company. For this we will be using a forecast dataset from Kaggle called Store Item Demand Forecast. You can download the dataset from `https://www.kaggle.com/c/demand-forecasting-kernels-only/data?select=train.csv`.

Once we have downloaded the dataset, we have to do some formatting on it. The first thing is that we will add an ID column to it. This should be the first column of the dataset. Next, make sure that the date field is in the format of YYYY-MM-DD; otherwise, Amazon Forecast will not accept it. Once that done, upload the dataset to S3 and then note its path.

Creating a Dataset Group

Now, open the Amazon Management Console and then search for *Amazon Forecast Service*. Click Create Dataset Group. The first step will be to give the dataset group a name and then choose the domain of forecasting. For our analysis, we have chosen a custom domain. See Figure 10-14.

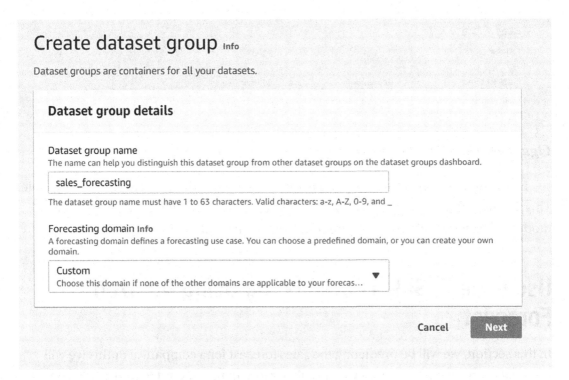

Figure 10-14. *Creating a dataset group name*

Defining Column Attributes

The next step is to give your dataset a name. Then, you have to define the column attributes. See Figure 10-15.

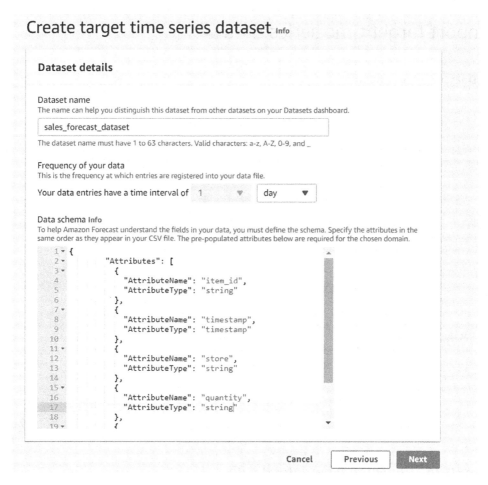

Figure 10-15. Defining column attributes

Remember, the attributes should be in the same sequence as in the dataset. Also, all the attributes besides the attributes that were already present in the JSON schema must be strings. That's why the store and quantity are mentioned as strings. You can define the interval as well. For us, it is a daily interval, so that's how it is set. Click Next.

Importing Data

The last step will be to start importing the data. Here you will provide the S3 path and the name to the dataset import. See Figure 10-16.

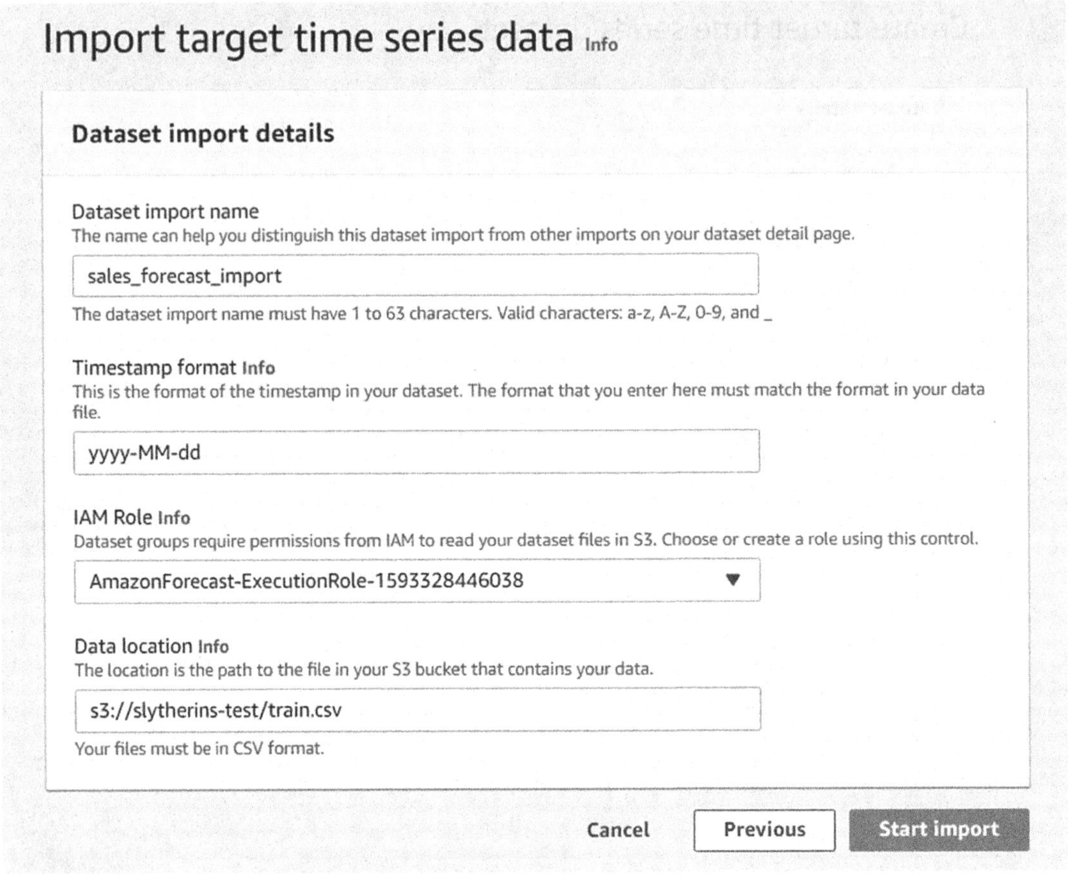

Import target time series data Info

Dataset import details

Dataset import name
The name can help you distinguish this dataset import from other imports on your dataset detail page.

sales_forecast_import

The dataset import name must have 1 to 63 characters. Valid characters: a-z, A-Z, 0-9, and _

Timestamp format Info
This is the format of the timestamp in your dataset. The format that you enter here must match the format in your data file.

yyyy-MM-dd

IAM Role Info
Dataset groups require permissions from IAM to read your dataset files in S3. Choose or create a role using this control.

AmazonForecast-ExecutionRole-1593328446038   ▼

Data location Info
The location is the path to the file in your S3 bucket that contains your data.

s3://slytherins-test/train.csv

Your files must be in CSV format.

Cancel   Previous   **Start import**

Figure 10-16. *Configuring the dataset path*

Making Predictions

Now, click "Start import." Once the import is done, we have to click "Start predictor training." Here we will inform Amazon Forecast about the type of forecast that we have to make and the horizon of forecasts. Figure 10-17 shows the inputs that we have to make.

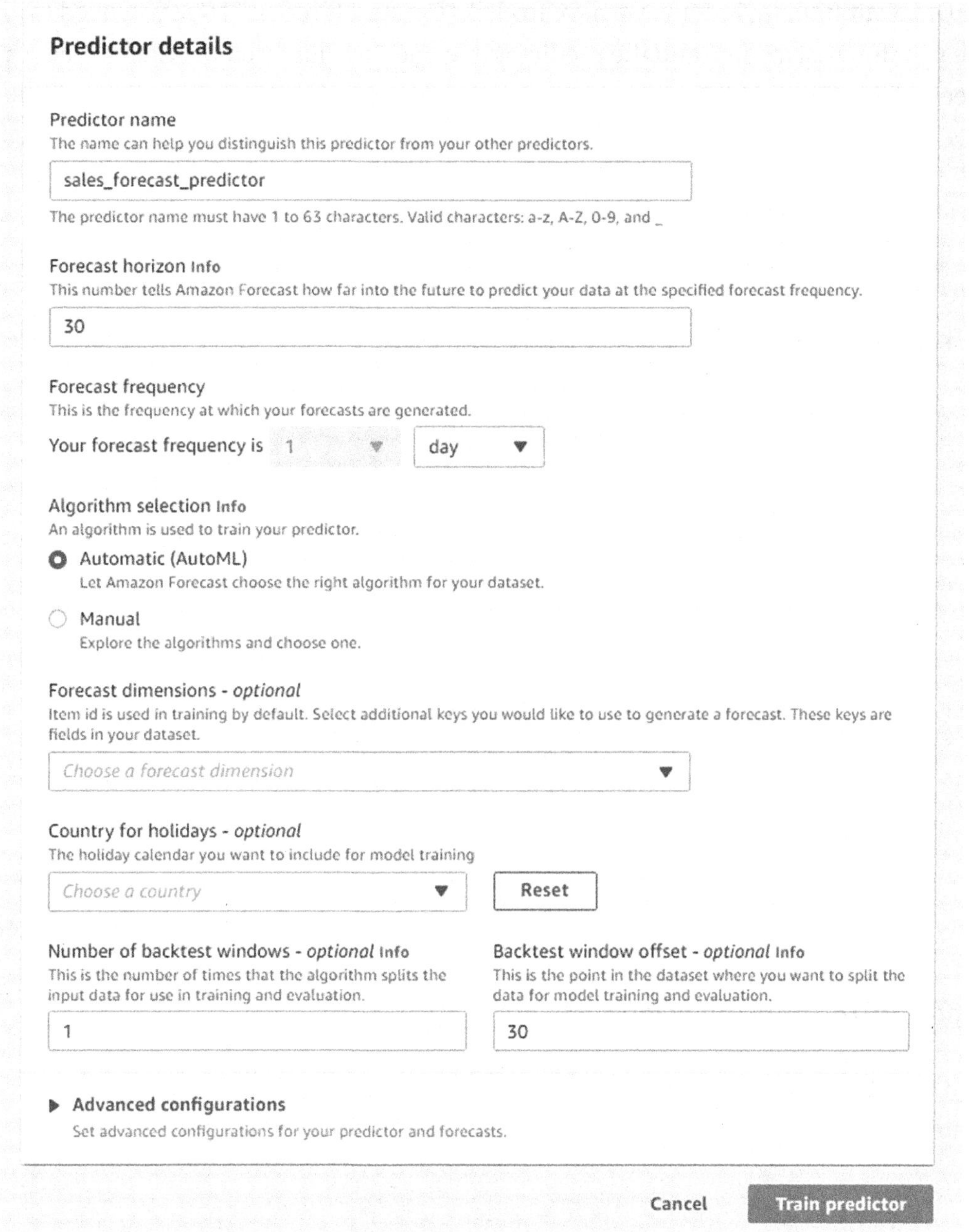

Predictor details

Predictor name
The name can help you distinguish this predictor from your other predictors.

sales_forecast_predictor

The predictor name must have 1 to 63 characters. Valid characters: a-z, A-Z, 0-9, and _

Forecast horizon Info
This number tells Amazon Forecast how far into the future to predict your data at the specified forecast frequency.

30

Forecast frequency
This is the frequency at which your forecasts are generated.

Your forecast frequency is   1   ▼   day   ▼

Algorithm selection Info
An algorithm is used to train your predictor.

○ **Automatic (AutoML)**
   Let Amazon Forecast choose the right algorithm for your dataset.

○ **Manual**
   Explore the algorithms and choose one.

Forecast dimensions - *optional*
Item id is used in training by default. Select additional keys you would like to use to generate a forecast. These keys are fields in your dataset.

Choose a forecast dimension   ▼

Country for holidays - *optional*
The holiday calendar you want to include for model training

Choose a country   ▼   Reset

Number of backtest windows - *optional* Info
This is the number of times that the algorithm splits the input data for use in training and evaluation.

1

Backtest window offset - *optional* Info
This is the point in the dataset where you want to split the data for model training and evaluation.

30

▶ **Advanced configurations**
   Set advanced configurations for your predictor and forecasts.

Cancel   **Train predictor**

Figure 10-17. *Inputs for starting the training job*

The training will start after this. Wait until the training finishes, and if you feel that AutoML is taking too much time, you can select any of the custom predictors. Go with the manual algorithms.

Once our training is done, we have to click "Create a forecast." Here you need to give your forecast a name and the predictor that you have just created. Click "Create a forecast." See Figure 10-18.

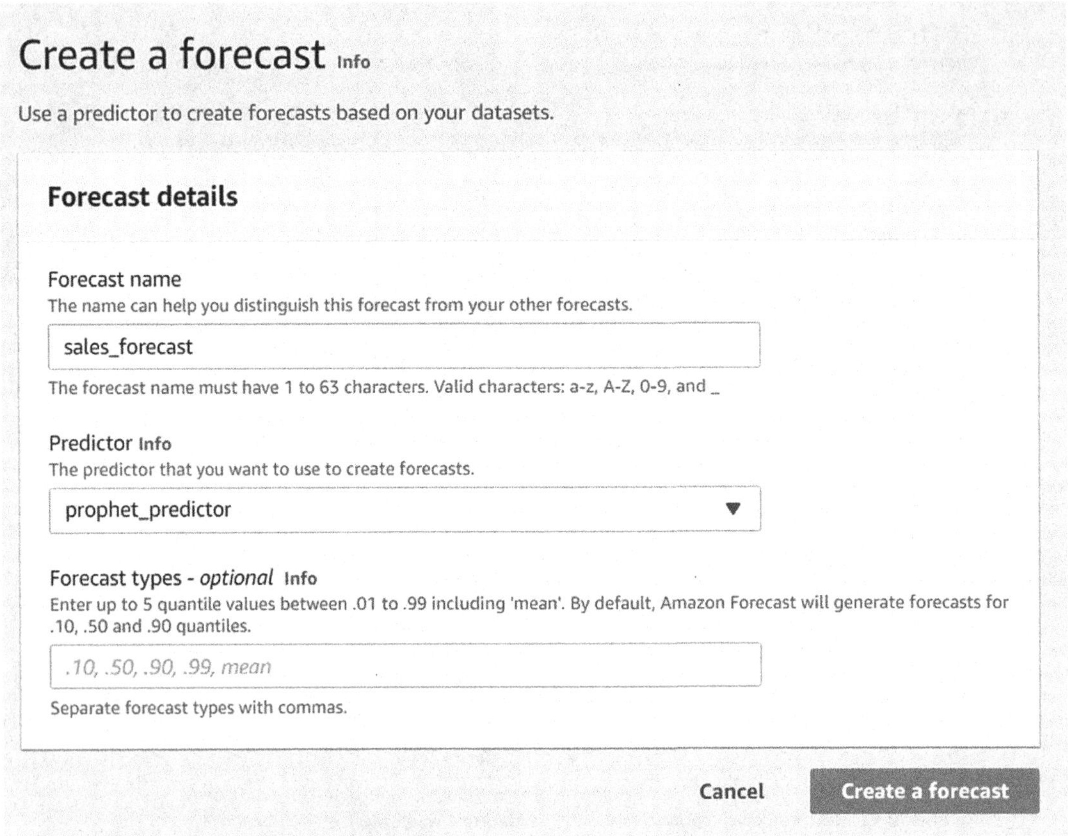

Create a forecast Info

Use a predictor to create forecasts based on your datasets.

Forecast details

Forecast name
The name can help you distinguish this forecast from your other forecasts.

sales_forecast

The forecast name must have 1 to 63 characters. Valid characters: a-z, A-Z, 0-9, and _

Predictor Info
The predictor that you want to use to create forecasts.

prophet_predictor   ▼

Forecast types - *optional* Info
Enter up to 5 quantile values between .01 to .99 including 'mean'. By default, Amazon Forecast will generate forecasts for .10, .50 and .90 quantiles.

.10, .50, .90, .99, mean

Separate forecast types with commas.

Cancel   Create a forecast

Figure 10-18. *Generating forecasts*

Once the forecast is done, you can click "Forecast Lookup" and view it. See Figure 10-19.

Forecast lookup Info

When you create a forecast, Amazon Forecast generates forecasts for each unique item in your target time-series dataset. Use the forecast lookup to find your forecasts.

Forecast details

Forecast	Start date Info	End date Info
Choose the forecast whose forecasts you want to view.	This is the start date for the historical demand you want to view.	This is the end date for the forecast that you want to view.
Choose a forecast name ▼	YYYY/MM/DD 🔲	YYYY/MM/DD 🔲
	00:00:00	00:00:00
	Use 24-hour format.	Use 24-hour format.

`Get Forecast`

Figure 10-19. *Forecast lookup*

Just give the horizon for forecasts and you will get a very nice visualization, as shown in Figure 10-20.

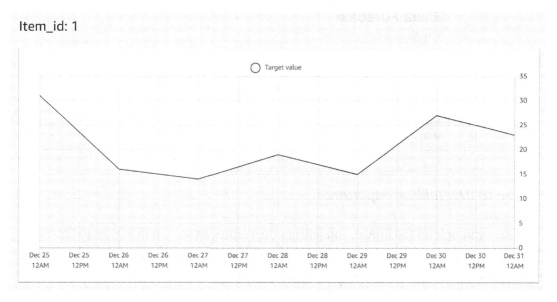

Figure 10-20. *Generated forecasts*

This finishes our discussion of using Amazon Forecast for making sales forecasts. As mentioned before, we can create different types of forecasts for different datasets. Let's now look at the last use case where we will use Amazon Textract to extract textual information from different file formats.

Use Case 3: Image Text Extraction Using Amazon Textract

Using Amazon Textract, not only can we extract the text from the images, PDFs, Word files, etc., but we can also extract the tabular and form-based data as well. Again, like with all the previous services, the process is simple. We just need to upload the image containing information, and the data will be extracted. Let's start the process by first going to the Amazon Management Console and searching for *Amazon Textract service*. Next, click "try Amazon Textract." Finally, click the upload document.

Extracting Tabular Information

First, let's apply Textract on some tabular data. For this we will be using Figure 10-21.

Sales Forecast

When are current deals closing?

https://medium.com/pipedrive-analytics/sales-forecast-report-cb56cdae0af5#.dw35bggul

	# Deals Closing	Expected Total	Weighted Total
January	34	$100,455	$24,109
February	30	$89,440	$21,466
March	35	$98,655	$23,677
April	28	$45,000	$10,800
May	24	$55,689	$13,365
June	25	$104,555	$25,093

Figure 10-21. *Table of importance*

Once you upload the image, you will get the results, as shown in Figure 10-22. The first extraction is of the raw text, while the second extraction is of the table. We can see that Textract worked perfectly, as shown in Figure 10-23.

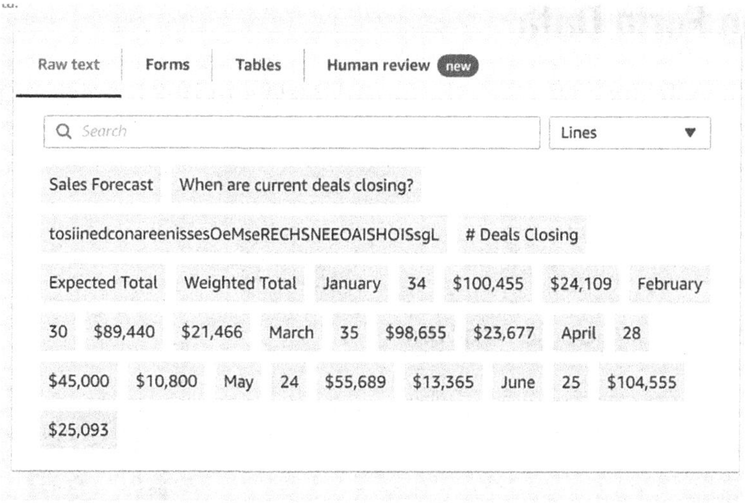

Figure 10-22. *Extracted keywords*

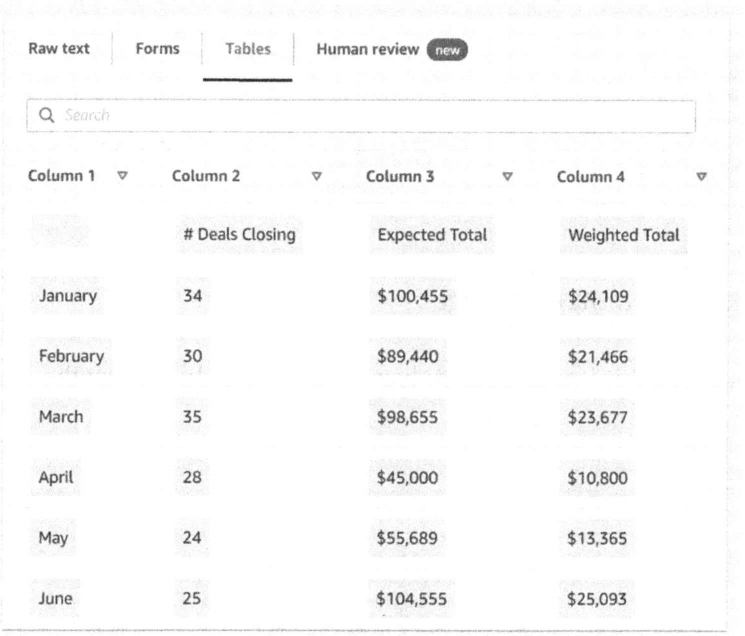

Figure 10-23. *Tabular keywords extracted*

Extracting Form Data

Now, let's see how Textract works on the form data. Figure 10-24 shows the image we will use.

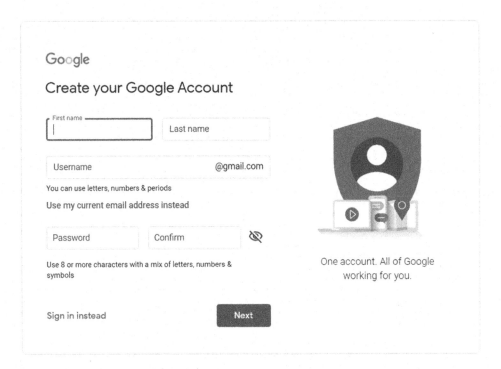

Figure 10-24. *Form data test image*

When we upload this image and analyze the form results, we get the results shown in Figure 10-25.

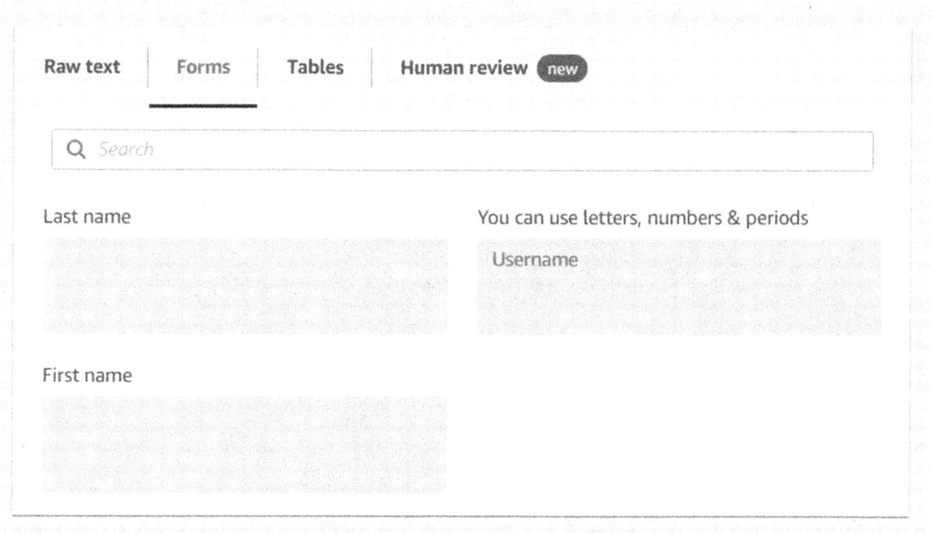

Figure 10-25. *Form data extracted*

You can see that most of the fields of the form are successfully extracted by Textract. Similarly, we can give PDF files and other supported formats to Textract, to get the required fields. We can connect to the Textract API as well so that directly the results can be absorbed by a scripting language like Python and further analysis can be made.

Conclusion

In this chapter, you learned about different Amazon services in the domain of machine learning. They are the ready-made solutions provided by Amazon that minimize the coding knowledge so that people with deep mathematics/statistical backgrounds can conduct their analysis. This concludes the last chapter of this book.

APPENDIX A

Creating a Root User Account to Access the Amazon Management Console

Follow these steps to create a root account to access the Amazon Management Console:

1. Go to `https://aws.amazon.com/console/`.

2. Click Create Free Account.

3. Enter an email, password, and name for your AWS account.

4. Select the type of account: Professional or Personal.

5. Fill in the details.

6. Give your credit/debit card details. Once you've done that, the account will be created. (You may have to verify your email address.)

7. Log in to the Amazon Management Console with the username and password you just created.

Now you are ready to use the Amazon Management Console interface.

227

© Himanshu Singh 2021
H. Singh, *Practical Machine Learning with AWS*, https://doi.org/10.1007/978-1-4842-6222-1

APPENDIX B

Creating an IAM Role

Follow these steps to create an IAM role:

1. Log in to the Amazon Management Console and search for *IAM service*.

2. Click the Roles section.

3. Click Create Role.

4. Search for *Go for SageMaker*.

5. Search for *Full Access permission* and then keep clicking Next.

6. Click Create Role.

© Himanshu Singh 2021
H. Singh, *Practical Machine Learning with AWS*, https://doi.org/10.1007/978-1-4842-6222-1

APPENDIX C

Creating an IAM User

Follow these steps to create an IAM user:

1. Log in to the Amazon Management Console and search for *IAM service*.

2. Click the Users section.

3. Click Add User.

4. Enter a username and select AWS Management Console Access.

5. Enter a password.

6. Click Next and search for a policy that the user can use. You can go for Admin or any particular policy. Let's select AmazonSageMakerFullAccess.

7. Keep clicking Next and then click Create User.

The user will be successfully created, and the next time you can log in to the Amazon Management Console using this username and password instead of the root user.

© Himanshu Singh 2021
H. Singh, *Practical Machine Learning with AWS*, https://doi.org/10.1007/978-1-4842-6222-1

Creating an S3 Bucket

Follow these steps to create an S3 bucket:

1. Log in to the Amazon Management Console and search for *S3 service.*

2. Click Create Bucket.

3. Give your bucket a name. Follow the rules (it should be DNS compliant).

4. Select the region of your choice.

5. Now click Next until you reach the permissions. Give the bucket public access so that you can use S3 buckets with other services. (Do not give public access if you have confidential information. In that case, you should go with policies.)

6. Create the S3 bucket and note the path.

Now you can access the S3 bucket at the URL you have just written down.

© Himanshu Singh 2021
H. Singh, *Practical Machine Learning with AWS*, https://doi.org/10.1007/978-1-4842-6222-1

APPENDIX E

Creating a SageMaker Notebook Instance

Follow these steps to create a SageMaker notebook instance:

1. Log in to the Amazon Management Console and search for *SageMaker*.

2. Go to the Notebook Instances section.

3. Click Create Notebook Instance.

4. Give the instance a name and select the type. If you want to use the free version, select ml.t2.medium; otherwise, you can select a paid version.

5. Select the SageMaker IAM role that you defined.

6. Click Create Notebook Instance. In a few minutes your instance will be ready.

7. Always remember to stop the instance when your code is done.

© Himanshu Singh 2021
H. Singh, *Practical Machine Learning with AWS*, https://doi.org/10.1007/978-1-4842-6222-1

Index

© Himanshu Singh 2021
H. Singh, *Practical Machine Learning with AWS*, https://doi.org/10.1007/978-1-4842-6222-1

Printed in the United States
By Bookmasters